AN INTRODUCTION
to MICROWAVE
MEASUREMENTS

AN INTRODUCTION
TO MICROWAVE
MEASUREMENTS

Ananjan Basu

CRC Press
Taylor & Francis Group
Boca Raton London New York

CRC Press is an imprint of the
Taylor & Francis Group, an **informa** business

CRC Press
Taylor & Francis Group
6000 Broken Sound Parkway NW, Suite 300
Boca Raton, FL 33487-2742

First issued in paperback 2017

ISBN-13: 978-1-4822-1435-2 (pbk)
ISBN-13: 978-1-138-74961-0 (hbk)

Library of Congress Cataloging-in-Publication Data

Basu, Ananjan.
    An introduction to microwave measurements / author, Ananjan Basu.
        pages cm
    Includes bibliographical references and index.
    ISBN 978-1-4822-1435-2 (hardback)
    1. Microwave measurements. 2. Radio measurements. I. Title.

    TK7876.B375 2015
    621.381'3--dc23   2014039022

Visit the Taylor & Francis Web site at
http://www.taylorandfrancis.com

and the CRC Press Web site at
http://www.crcpress.com

Contents

Preface

This book is the outcome of teaching a master's-level course on microwave measurements at the Indian Institute of Technology Delhi. While teaching these topics, I realized that there are no suitable textbooks on this subject—only advanced reference books, which are not suitable for students with little or no background in microwave engineering. Accordingly, this book has been written to ensure that the reader is in a position to understand more advanced books and papers on the subject. A fairly wide range of topics have been addressed, but naturally the depth has been kept small. At the same time, care has been taken to ensure that some meaningful introduction to the latest instruments and techniques is given.

The book is oriented around the most common instruments used in microwave measurements—the network analyzer, the spectrum analyzer, and synthesized microwave sources. It is hoped that the reader will gain some familiarity with these instruments and understand what goes on inside them.

Acknowledgments

I am grateful to Gowrish B, ex-student of the Centre for Applied Research in Electronics, IIT Delhi, for compiling much of the material related to oscilloscopes and wafer probing. I am also grateful to our students, particularly Ritabrata Bhattacharya, Srujana Kagita, and Sweta Agarwal, for assistance with writing this book. I am grateful to my colleagues Prof. S.K. Koul, Dr. Mahesh Abegaonkar, and Dr. Karun Rawat at CARE IIT Delhi for their support. Finally, I thank my wife and son for the patience they have shown when I was busy meeting the deadline for finishing the book.

About the Author

Ananjan Basu completed his BTech in electrical engineering and MTech in communication and radar engineering from IIT Delhi in 1991 and 1993, respectively, and a PhD in electrical engineering from the University of California, Los Angeles, in 1998. He has been employed at the Centre for Applied Research in Electronics, IIT Delhi as visiting faculty from 1999 to 2000, as assistant professor from 2000 to 2005, as associate professor from 2005 to 2012, and as professor from 2013. His specialization is in microwave and millimeter-wave component design and characterization. He has been instructing students in these areas since 1999. His research interests include microwave component development, active and reconfigurable antennas, and investigation of magnetic materials such as low temperature co-fired ceramics (LTCC)-ferrite and ferromagnetic nanowires.

Professor Basu has published more than 80 papers in journals and conferences and has supervised or co-supervised 4 PhD and more than 50 master's dissertations. Some of the awards that Dr. Basu has received, jointly with his colleagues and students, are the Indian National Academy of Engineering Award for Best Undergraduate Project 2001; Indian National Academy of Engineering Award for Best Postgraduate Project 2003; IEEE-MTTS UG Scholarship 2006; Best Student Paper Award, IEEE International Conference on Antennas Propagation and Systems, Malaysia, December 2009; and the M.N. Saha Award for Best Application-Oriented Paper in the *Journal of IETE*, India, 2013.

Dr. Basu has also undertaken several sponsored projects on the development of microwave systems, such as wide-band radio frequency (RF) radiation detection and microwave imagers. His research activities in the recent past have included ultra-wide-band pulse generation and transmission, active and reconfigurable antennas, and dielectric waveguides.

Dr. Basu is a member of the Institute of Electrical and Electronics Engineers (IEEE) and has served as chairman of the IEEE-MTTS Delhi Chapter. He also serves as a subject expert in several committees of the government of India.

1

Introduction

1.1 Aim and Scope

This book has been written primarily for senior undergraduate and first-year master's-level students who intend to gain some degree of proficiency in subjects related to radio frequency (RF) and microwave systems. It is also likely to be found useful by professionals in the wireless industry who deal with such systems but are not specialists in the area.

As the title implies, this is a textbook; hence, the topics discussed are mostly well-established, mature subjects, although a few modern innovations have been described in Chapters 8 and 9. Since the area of microwave measurements is not new, there are many references that discuss these topics in depth. However, it is not easy for beginners to benefit from these books—almost all are aimed at specialists in the field. The area of traditional microwave engineering has seen a steady decrease in popularity among students in the last few decades, despite the proliferation of wireless systems today. Consequently, very few students who graduate in electrical engineering today have a knowledge of this subject beyond basic distributed circuit analysis. With this limited background, it is difficult to follow references such as Dunsmore (2012), Engen (1992), Chen et al. (2004), Sucher and Fox (1963), or Rohde (1997). Books like Sucher and Fox (1963) or Ginzton (1957) have also become outdated.

While teaching the subject of microwave measurements to first-year postgraduate students, I have always felt the lack of a textbook that would be of use to them. With the proliferation of in-depth technical information on the Internet today, the need is more for a simplified explanation of the techniques used today and not a compendium of the latest information—for the advanced student, there is no dearth of technical literature. One goal of this book is to equip the reader to grasp the material in more specialized references in the area of modern microwave measurements.

With this aim, the book has been written in a way that is different from many textbooks. A large part of modern microwave measurements (excluding antenna measurements) has crystallized around a few instruments—the network analyzer, spectrum analyzer, synthesized source, and oscilloscope.

These instruments have become extremely sophisticated in recent times, and consequently, a large part of this book has been written in a "how it works" spirit, strongly borrowing from instrument catalogs and application notes. At the same time, traditional introductory topics in measurement, such as the definitions of *resolution* and *accuracy* or the importance and determination of measurement uncertainties, are omitted; these are well detailed in many other textbooks. Another important part of measurements today is processing the data from these instruments, leading to device and component modeling. This area, however, is only hinted at in this book, as it is a more advanced topic. In the same spirit, the book gives a first-level description of many topics and then refers the reader to advanced references.

Examples involving numerical values are scattered throughout the book—it has been observed that students generally grasp the subject better when actual numbers are given, rather than symbolic relations.

1.2 General Electronic Measurements and Frequency Limitations

Students of most engineering disciplines get acquainted with the oscilloscope (the cathode-ray oscilloscope [CRO] is still seen in many educational laboratories) early in their studies. This instrument, along with the waveform generator, forms the basis of laboratory measurements at lower frequencies—almost any measurement can be carried out with them (and a few accessories or components such as op-amps). As is well known, this cannot work beyond ~100 MHz, when the wavelength is no longer very large compared to the system dimensions (for example, a 1-m-long co-axial cable is only one-third wavelength at 100 MHz). This of course introduces distributed circuits, which can be analyzed reasonably well using transmission line theory. Circuit analysis and design becomes mathematically tedious, but no fundamental changes are required up to much higher frequencies, provided well-known precautions are taken. Measurement procedures, however, change drastically—instead of terms like *node voltage*, *impedance*, or *voltage gain*, we bring in terms like S_{21}, *harmonic level*, or *dBc*. The primary reason behind this is the near impossibility of measuring (or even defining) *voltage* at a particular point. In contrast, it is easy to sample the voltage waves propagating in two directions along a transmission line—moreover, these waves can actually be separately sampled, and adequate circuit characterization can be carried out from the sampled values. It is not that *impedance* or *voltage* gain cannot be defined at higher frequencies—they are simply difficult to measure directly. If required, they can always be calculated from

S-parameter measurement. This is the goal of the network analyzer, and this instrument is discussed in Chapters 3 and 4.

The spectrum of a signal, of course, is important from audio to optical frequencies. At the lower frequencies, it is conveniently measured by carrying out mathematical operations on data collected through an oscilloscope. At higher frequencies, the specialized spectrum analyzer is used—the operation of this instrument is described in Chapter 5.

Electrical noise is a phenomenon that becomes more and more disturbing as we climb in frequency. Actually, it is high bandwidth that brings in high noise power, but high bandwidths (say, >1 GHz) are naturally encountered only at microwave frequencies. An introductory description of measuring noise figures is given in Chapter 6.

Another term that is frequently encountered in the wireless industry is *phase noise*. Unfortunately, very few nonspecialists know how this is quantified, although they install, operate, and maintain wireless equipment for which phase noise is a key design parameter. A simplified way to understand this subject, and some techniques for measuring phase noise are given in Chapter 6.

Obviously, any measurement requires a stimulus—at low frequencies we have waveform generators, while at microwave frequencies there are several types of microwave signal generators. A few of the most commonly used generators are discussed in Chapter 7. Some other types of sources (Gunn diodes, electron tubes, ultra-wide band (UWB) pulse generators, and waveform generators for microwave frequencies) that are less important in metrology are omitted. Of course, the discussion is restricted to an overview; mathematical analysis of synthesized sources (phase-locked loop [PLL] or direct digital synthesis [DDS]) is described in detail in many reference books.

As mentioned above, microwave measurements are motivated by the fact that measuring node voltage without disturbing the circuit is difficult at microwave frequencies. Historically, that is indeed the motivation for the development of a distinct approach to measurements at high frequencies. Today, however, technology has come full circle in a sense, and oscilloscopes that can actually measure node voltage without disturbing the circuit have been developed that work beyond 30 GHz. Some aspects of this technology are discussed in Chapter 8.

The last topic, wafer probing, is discussed in Chapter 9. Today it is possible to probe an integrated circuit (IC) while it is part of a semiconductor wafer, that too at frequencies exceeding 100 GHz. An overview of this technology is given.

Some mathematical tools and also a description of the basic circuit elements that are used to realize instruments are given in Chapter 2. Some readers may prefer to skip this chapter, as much of the subsequent material can be understood without this background.

One important topic that is not discussed here is antenna measurement. Traditionally, this is treated separately from the circuit-oriented measurements described here. There is specialized literature available treating antenna measurements—IEEE (1980) in particular is a good starting point.

1.3 Applications and Importance of Microwave Measurements

In addition to the traditional areas of radar and military technologies, microwave technologies are now indispensable in almost all areas of electronics. Cellular systems obviously have microwave transmitters and receivers at their core, but even the microprocessor today operates with clock speeds of >3 GHz, necessitating distributed circuit concepts in designing the circuit boards inside a computer. Consequently, characterization of many components and systems that previously relied on low-frequency (or lumped-circuit) concepts today requires the network analyzer or spectrum analyzer. The importance of microwave measurements is thus too obvious to require a detailed listing of specific applications. I will only point out two specific uses of microwave instruments that are critical in modern technology development:

1. Because of the severe demand on spectrum for commercial wireless systems, it has become essential to reduce the guard bands between channels to the bare minimum. This minimum is decided by the phase noise in the carrier and by the selectivity of band-pass filters. Techniques exist for accurate measurement of both, as we shall see.

2. **Electromagnetic interference** (EMI)/**electromagnetic compatibility** (EMC) measurements are critical today as mandated by multiple internationally accepted standards. Handheld spectrum and network analyzers have come to the market in the past few years for such measurements in the field.

1.4 Overview of State-of-the-Art Microwave Measurements

State-of-the-art capabilities of common modern microwave instruments include upper frequency limit, frequency resolution, noise floor, amplitude accuracy, etc. A few instruments that show impressive performance, and yet are quite affordable, are shown here. Figure 1.1 shows a 40 GHz spectrum analyzer, with up to 1 Hz resolution bandwidth and displayed noise level below –130 dBm above 1 MHz.

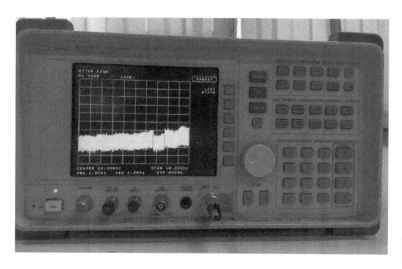

FIGURE 1.1
A 40 GHz spectrum analyzer.

A vector network analyzer that works from 10 MHz to 67 GHz, and includes a free-space material property measurement facility for the 40–60 GHz range, is shown in Figure 1.2. Noise floor is below –120 dB, and it can measure S-parameters with a magnitude accuracy better than 0.1 dB and a phase accuracy better than 1°. This also contains an internal signal generator synthesizing an output up to 60 GHz with sub-Hz resolution.

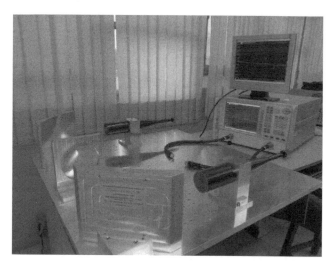

FIGURE 1.2
Vector network analyzer (VNA) with material measurement setup.

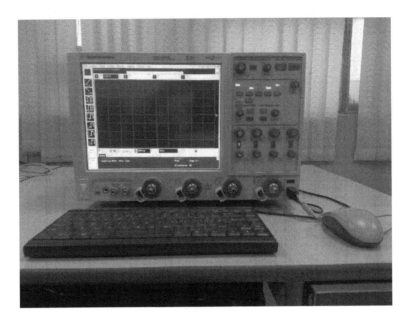

FIGURE 1.3
A digital storage oscilloscope up to 25 GHz.

A digital storage oscilloscope that displays up to four independent real-time waveforms with frequencies up to 25 GHz is shown in Figure 1.3.

The Internet, of course, will provide a wealth of the latest information.

References

Chen, L., C.P. Neo, C.K. Ong, V.V. Varadan, and V.K. Varadan. *Microwave Electronics: Measurement and Materials Characterization*. Wiley, Chichester, U.K., 2004.

Dunsmore, J.P. *Handbook of Microwave Component Measurements*. John Wiley & Sons, Chichester, U.K., 2012.

Engen, G. *Microwave Circuit Theory and Foundations of Microwave Metrology*, Peter Peregrinus, Ltd., London, U.K., 1992.

Ginzton, E.L. *Microwave Measurements*, McGraw-Hill, New York, 1957.

IEEE. *Standard Test Procedures for Antennas*. ANSI/IEEE Standard 149-1979. New York, 1980.

Rohde, U.L. *Microwave and Wireless Synthesizers: Theory and Design*, John Wiley & Sons, New York, 1997.

Sucher, M. and J. Fox, Eds. *Handbook of Microwave Measurements*. Vol. 1, 3rd Brooklyn, NY: Polytechnic Press, 1963.

2

Background Information

In this chapter an introduction to the mathematical basis behind microwave measurements, and also an overview of the components used in microwave instruments, is given.

The relevant mathematical techniques are essentially: (1) S-parameters and related black-box representation techniques and (2) spectra of commonly encountered signals.

2.1 S-Parameters and Related Black-Box Representation

It is well known that linear time-invariant circuits (radio frequency (RF)/ microwave or not) can be completely described by any of the multiport representations, such as S-parameters, Z-parameters, etc. The theory behind such descriptions is detailed in many references, such as Pozar (1998). Here we will review a few aspects of black-box representations that are important from the measurement point of view. A good introduction to basic theory behind S-parameters is given in Pozar (1998) and other references—it is assumed that the reader is familiar with this material.

At frequencies greater than ~100 MHz, the preferred approach to black-box representation is the use of S-parameters. There are two main reasons for this. First, the S-parameters are the ones that are easy to measure directly; details about such measurement are given in Chapter 4. Second, these are the parameters that give useful information about the properties of components that are commonly used at the higher frequencies. Components used at lower frequencies (e.g., transformer) may naturally have a more appropriate description in terms of some other multiport parameters.

For example, consider a low-pass filter with cutoff frequency 1 GHz. Typical S-parameters of such a filter at two frequencies (0.9 and 1.3 GHz) are

$$\begin{bmatrix} 0.22\angle 39° & 0.98\angle 129° \\ 0.98\angle 129° & 0.22\angle 39° \end{bmatrix} \text{ at } 0.9 \text{ GHz}$$

and

$$\begin{bmatrix} 0.99\angle 72° & 0.13\angle -17° \\ 0.13\angle -17° & 0.99\angle 72° \end{bmatrix} \text{ at 1.3 GHz}$$

A cursory inspection shows:

1. The off-diagonal terms are close to 1 in magnitude below the cutoff frequency and are well below 1 above the cutoff frequency.
2. The reverse is true for the diagonal elements.

Since the magnitudes of the diagonal terms represent reflection, and the magnitudes of the off-diagonal terms represent transmission, the above observation agrees with what we expect from a low-pass filter.

Now consider the Z-parameters of the same filter at the same frequencies. These are

$$j\begin{bmatrix} -56 & 87 \\ 87 & -56 \end{bmatrix} \text{ at 0.9 GHz}$$

and

$$j\begin{bmatrix} 69 & 9.6 \\ 9.6 & 69 \end{bmatrix} \text{ at 1.3 GHz}$$

It is not easy to relate these numbers to any aspect of filter performance without further computations.

2.1.1 S-Parameters

We will use the following definition of S-parameters (there is some variation in this among Pozar, Ha, and others (1998)).

Let us consider an N-port circuit. The definition (following Ha (1981)) will start by assuming that voltages and currents can be defined at the ports. This assumption becomes questionable (but can be justified) when the ports are rectangular metal waveguides, and cannot be justified (except from an abstract mathematical point of view) when the ports are optical fibers (obviously there is no current flow in an optical fiber that does not contain a conductor). A different and more general definition of S-parameters, starting from electric and magnetic fields, has to be used in such cases (Itoh (1989)).

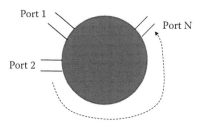

FIGURE 2.1
S-parameters of an *N*-port circuit.

Let the voltages and currents at the ports be denoted: $V_1, V_2, \ldots, V_N$ and $I_1, I_2, \ldots, I_N$. Using the usual frequency domain or phasor notation, these are complex numbers that depend on frequency.

To define them unambiguously, the direction convention has to be specified. This is shown in Figure 2.2, where + and − signs are used for voltages and arrows are used for currents.

The interpretation of these symbols is

1. The voltage is (potential on the terminal marked +) − (potential on the terminal marked −). The choice of + and − is *in principle arbitrary and at the user's discretion*, but in practical circuits, as we shall see soon, a convenient choice is obvious.

2. The current direction is as shown by the arrows. Current is taken to be flowing *into the circuit* on the terminal marked +. Notice that having selected the terminals marked + for the voltages, there is no further choice for the current. For circuits of interest to RF/microwave measurements, there is an exactly equal current flowing out of the circuit on the terminal marked −.

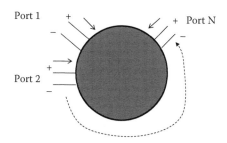

FIGURE 2.2
Voltage and current directions for the *N*-port.

Now the mathematical definition of S-parameters is simple. Define the variables (complex) $a_1, \ldots, a_N$ and $b_1, \ldots, b_N$ by

$$a_i = (V_i + R\,I_i)/(2\sqrt{R}) \quad \text{and} \quad b_i = (V_i - R\,I_i)/(2\sqrt{R}) \text{ for } i = 1, \ldots, N \qquad (2.1)$$

Then the $N \times N$ S-matrix is defined through the relation

$$\begin{bmatrix} b_1 \\ \vdots \\ b_N \end{bmatrix} = \begin{bmatrix} S_{11} & \cdots & S_{1N} \\ \vdots & \ddots & \vdots \\ S_{N1} & \cdots & S_{NN} \end{bmatrix} \begin{bmatrix} a_1 \\ \vdots \\ a_N \end{bmatrix} \qquad (2.2)$$

A symbol R has been used here. This is an *arbitrary* real positive number. Very often Z_0 or Z_c is used in place of R. Obviously, the S-parameters (elements of the S-matrix) are dependent on R. The terminology commonly used is that the S-parameters are normalized to R or with a reference R. Note that R has dimensions of resistance.

Example 2.1

Let us calculate the S-parameters of a microstrip T-junction at very low frequencies (say, 1 KHz). This is shown in Figure 2.3.
    S-parameters normalized to some R (typically 50 Ω) are required. *Assumption*: The whole circuit is much smaller than a wavelength. Start from the definition:

$$\begin{bmatrix} b_1 \\ b_2 \\ b_3 \end{bmatrix} = \begin{bmatrix} S_{11} S_{12} S_{13} \\ S_{21} S_{22} S_{23} \\ S_{31} S_{32} S_{33} \end{bmatrix} \begin{bmatrix} a_1 \\ a_2 \\ a_3 \end{bmatrix}$$

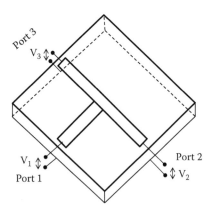

FIGURE 2.3
A microstrip T-junction.

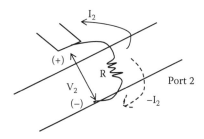

FIGURE 2.4
Zooming in on port 2 of the *T*-junction.

At this stage there is no complete circuit. We need to connect something to the ports. Also, there should be a source connected; otherwise, all voltages and currents become 0, which is of no use.

Now if we could somehow make these connections so that a_2 and a_3 become 0, we get a simple equation from the first line of the matrix: $b_1 = S_{11} a_1 \Rightarrow S_{11} = \frac{b_1}{a_1}$.

This implies that S_{11} may be evaluated if V_1 and I_1 are known.

To make $a_2 = 0$, note that $a_2 = (V_2 + R\, I_2)/2\sqrt{R}$. So, if $V_2/(-I_2) = R$, we will get the desired condition satisfied.

Ensuring that $V_2/(-I_2) = R$ is easy. We simply connect a resistor R at port 2. Zooming in on port 2, we have what is shown in Figure 2.4. (Remember, I_2 has to flow *into* the circuit, as per the agreed convention.)

Using Ohm's law, indeed, we have $\frac{V_2}{(-I_2)} = R$. Similarly, connecting a resistor R at port 3 will ensure that $a_3 = 0$. Finally, we have to connect a source at port 1. We can connect a voltage source or a current source, or a voltage source with source series impedance or some other network, including a source—the possibilities are many.

Let us connect a current source I_0. The circuit now is as in Figure 2.5.

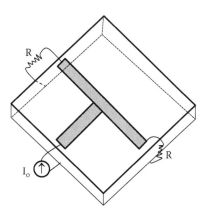

FIGURE 2.5
The complete circuit for evaluating *S*-parameters of the *T*-junction.

Since the circuit dimensions << wavelength, the entire *T*-junction is a simple node, and the simple rules of lumped circuits apply. The two resistors appear in parallel, and we get

$$V_1 = V_2 = V_3 = I_0 R/2$$

and

$$I_1 = I_0$$

So,

$$a_1 = \frac{V_1 + RI_1}{2\sqrt{R}} = \frac{I_0 R/2 + I_0 R}{2\sqrt{R}} = -\left(\frac{3\sqrt{R}}{4}I_0\right)$$

and

$$b_1 = \frac{V_1 + RI_1}{2\sqrt{R}} = \frac{I_0 R/2 - I_0 R}{2\sqrt{R}} = \left(\frac{\sqrt{R}}{4}I_0\right)$$

finally giving us $S_{11} = -1/3$.

Notice that in this very special case, the *S*-parameters are independent of *R*.

The interpretation of a_i and b_i as forward and backward traveling waves (strictly speaking, power waves [Ha (1981)]) is also well known but is worth repeating here; as the next example will show, this interpretation has to be used carefully.

Let us examine port 2 from Figure 2.2 as shown in Figure 2.6 and assume that a lossless transmission of *characteristic impedance R* is connected here.

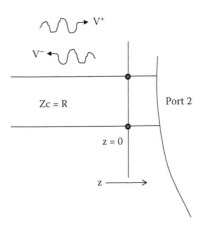

FIGURE 2.6
Voltage waves at port 2.

The transmission line is assumed to be oriented along the z-direction, with $z = 0$ defining the exact location of port 2. What is connected to the left of the transmission line is not important as long as there is a source connected to the circuit somewhere (or to the transmission line) that gives rise to nonzero voltage and current on the transmission line.

At $z = 0$, we have the well-known equations from transmission line theory (Pozar (1998)):

$$V(z = 0) = V^+(z = 0) + V^-(z = 0) \quad \text{and} \quad I(z = 0) = [V^+(z = 0) - V^-(z = 0)]/R \quad (2.3)$$

Adding and subtracting these equations, we get

$$a_2 = V^+(z = 0)/\sqrt{R} \quad (2.4)$$

and

$$b_2 = V^-(z = 0)/\sqrt{R} \quad (2.5)$$

This is obviously true for any of the ports. So, ignoring the factor $\sqrt{R}$, we can conclude:

1. a_i is the incoming voltage wave at port i.
2. b_i is the outgoing voltage wave at port i.

So, the [S] matrix relates the outgoing waves to the incoming waves, as is well known. Also,

$$S_{ij} = b_i/a_j \text{ provided } a_j \text{ is the only nonzero variable among } a_1, \ldots, a_N \quad (2.6)$$

However, confusion stems from the values of the voltages and currents evaluated at $z = 0$. Are these the same just to the left or right of $z = 0$? What happens if there is a physical discontinuity in the structure at $z = 0$? The next example is related to these questions.

Example 2.2

Let us calculate the S-parameters at 1 GHz of a lossless transmission line with characteristic impedance 50 Ω and length 10 cm. We will take R to be 10 Ω and assume that transmission electron microscope (TEM) waves propagate with velocity $= c = 3 \times 10^8$ m/s on this line.

Approach 1: Using the ABCD matrix and well-known conversion formulae (Pozar (1998)) we get

$$[S] = \begin{bmatrix} 0.9\angle - 12.5° & 0.43\angle - 102.5° \\ 0.43\angle - 102.5° & 0.9\angle - 12.5° \end{bmatrix}$$

Approach 2: Direct circuit analysis. From the defining equation,

$$S_{11} = \frac{b_1}{a_1}\Big|_{a_2=0} \quad \text{and} \quad S_{21} = \frac{b_2}{a_1}\Big|_{a_2=0}$$

Now

$$a_2 = 0 \quad \frac{V_2 + RI_2}{2\sqrt{R}} = 0$$

$$\Rightarrow V_2 = -R\,I_2$$

Recall that I_2 is defined as flowing *into* port 2.

So, connecting a resistance R at port 2 will ensure that $V_2 = -RI_2$, which in turn will ensure that $a_2 = 0$. A signal source has to be connected a port 1. Let this be a current source I_0 (many alternative sources can be used).

We now get the circuit shown in Figure 2.7.

Let Z_m be the impedance seen by the current source.

$$Z_m = Z_c \frac{R + jZ_c \tan \beta L}{Z_c + jR \tan \beta L} = 50 \frac{10 + j50 \tan \beta L}{50 + j10 \tan \beta L}$$

So, $V_1 = I_0 Z_m$ and $I_1 = I_0$.

S_{11} can be evaluated from these relations, but for S_{21} we need V_2 and I_2 as well. For this, we use the ABCD matrix of the transmission line.

$$\begin{bmatrix} V_1 \\ I_1 \end{bmatrix} = \begin{bmatrix} \cos \beta L & jZ_c \sin \beta L \\ \dfrac{j}{Z_c} \sin \beta L & \cos \beta L \end{bmatrix} \begin{bmatrix} V_2 \\ -I_2 \end{bmatrix}$$

Inverting the matrix,

$$\begin{bmatrix} V_2 \\ -I_2 \end{bmatrix} = \begin{bmatrix} \cos \beta L & -jZ_c \sin \beta L \\ \dfrac{-j}{Z_c} \sin \beta L & \cos \beta L \end{bmatrix} \begin{bmatrix} V_1 \\ I_1 \end{bmatrix}$$

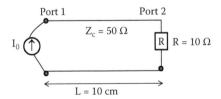

FIGURE 2.7
Circuit for calculating S-parameters of a transmission line.

So, $V_2 = \cos\beta L \; V_1 - jZ_c \sin\beta L \; I_1$ and $I_2 = V_1 \, (j/Z_c)\sin\beta L - I_1 \cos\beta L$. Using the expressions for V_1 and I_1 above,

$$V_2 = I_0[Z_m \cos\beta L - jZ_c \sin\beta L]$$

$$I_2 = I_0\left[j \, \frac{Z_m}{Z_c} \sin\beta L - \cos\beta L \right]$$

$$\therefore a_1 = \frac{I_0 Z_m + RI_0}{2\sqrt{R}}, \quad b_1 = \frac{I_0 Z_m - RI_0}{2\sqrt{R}}$$

which gives

$$S_{11} = \frac{Z_m - R}{Z_m + R}$$

$$b_2 = \frac{V_2 - RI_2}{2\sqrt{R}} = \frac{I_0}{2\sqrt{R}}\left[Z_m \cos\beta L - jZ_c \sin\beta L - j \, \frac{RZ_m}{Z_c} \sin\beta L + R\cos\beta L \right]$$

$$\therefore S_{21} = \frac{(Z_m + R)\cos\beta L - j\left(\dfrac{Z_c^2 + RZ_m}{Z_c} \right)\sin\beta L}{Z_m + R}$$

$$= \cos\beta L - j \, \frac{Z_c^2 + RZ_m}{RZ_c + Z_c Z_m} \sin\beta L$$

$$= \cos\beta L - j \, \frac{Z_c^2 + RZ_c \dfrac{R + jZ_c \tan\beta L}{Z_c + jR \tan\beta L}}{RZ_c + Z_c^2 \dfrac{R + jZ_c \tan\beta L}{Z_c + jR \tan\beta L}} \sin\beta L$$

$$= \cos\beta L - j \, \frac{Z_c^3 + jRZ_c^2 \tan\beta L + R^2 Z_c + jRZ_c^2 \tan\beta L}{RZ_c^2 + jR^2 Z_c \tan\beta L + RZ_c^2 + jZ_c^3 \tan\beta L} \sin\beta L$$

$$= \cos\beta L - j \, \frac{Z_c^2 + R^2 + 2jRZ_c \tan\beta L}{2RZ_c + j\left(R^2 + Z_c^2\right) \tan\beta L} \sin\beta L$$

$$= \frac{2RZ_c \cos\beta L + j\left(R^2 + Z_c^2\right)\sin\beta L - j\left(R^2 + Z_c^2\right)\sin\beta L + 2RZ_c \dfrac{\sin^2\beta L}{\cos\beta L}}{2RZ_c + j\left(R^2 + Z_c^2\right)\tan\beta L}$$

$$= \frac{2RZ_c/\cos\beta L}{2RZ_c + j\left(R^2 + Z_c^2\right)\tan\beta L}$$

$$= \frac{2}{2\cos\beta L + j\left(\dfrac{R}{Z_c} + \dfrac{Z_c}{R}\right)\sin\beta L}$$

Also, S_{11} can be simplified further:

$$S_{11} = \frac{Z_c \dfrac{R + jZ_c \tan\beta L}{Z_c + jR\tan\beta L} - R}{Z_c \dfrac{R + jZ_c \tan\beta L}{Z_c + jR\tan\beta L} + R}$$

$$= \frac{Z_c R + jZ_c^2 \tan\beta L - Z_c R - jR^2 \tan\beta L}{Z_c R + jZ_c^2 \tan\beta L + Z_c R + jR^2 \tan\beta L}$$

$$= \frac{j\left(Z_c^2 - R^2\right)\tan\beta L}{2Z_c R + j\left(Z_c^2 + R^2\right)\tan\beta L}$$

$$= \frac{j\left(\dfrac{Z_c}{R} - \dfrac{R}{Z_c}\right)\sin\beta L}{2\cos\beta L + j\left(\dfrac{Z_c}{R} + \dfrac{R}{Z_c}\right)\sin\beta L}$$

By symmetry, $S_{11} = S_{22}$ and $S_{21} = S_{12}$.
Finally, $\beta = \frac{2\pi}{\lambda} = \frac{2\pi f}{c} = \frac{2\pi \times 10^9}{3\times 10^8}$ and $L = 0.1$ m. So, $\beta L = 2.094$.
Evaluating the above expressions,

$$[S] = \begin{bmatrix} 0.9\angle -12.5° & 0.43\angle -102.5° \\ 0.43\angle -102.5° & 0.9\angle -12.5° \end{bmatrix}$$

If we try to interpret the mathematical steps in this example, we notice that to evaluate S_{11}, we should ensure that $a_2 = 0$. This is ensured by connecting a 10 Ω resistor at port 2. But a 50 Ω transmission line terminated with a 10 Ω resistor definitely gives reflected waves (reflection coefficient $\Gamma = -2/3$). At the end of the transmission line (around port 2), there are indeed voltage and current waves traveling in both directions. So, the physical picture of terminating the ports to set all a_j but one to 0 looks dubious. Intuitively, we expect to connect a 50 Ω termination to stop reflections, but that is contradictory to the derivation (which is well established and known to be correct).

To resolve this apparent contradiction, we should examine Figure 2.6 more closely. We can see that a_2 and b_2 are waves on an *external* transmission line of characteristic impedance R (in this case 10 Ω) connected to port 2. This transmission line is *not* part of the circuit. To set $a_2 = 0$ on this transmission line, we will naturally terminate it by 10 Ω. A transmission line of 10 Ω characteristic impedance, terminated by a 10 Ω resistor at one end, looks like a 10 Ω resistor from the other end. So, the 10 Ω termination used in the derivation may be regarded as a 10 Ω transmission line, terminated by a 10 Ω resistor, on which naturally there is no reflection.

During analysis, this extra transmission line is usually omitted since it plays no role in the mathematics. We are, however, left with a_2 and b_2 at port 2, when only a resistor is connected at port 2. In that case, assigning a physical meaning to such terms as *waves* in a lumped resistor is not meaningful, and these terms are best left as mathematical entities. Notice, however, that all this rests on the assumption that voltages and currents are well defined. As mentioned earlier, there are indeed situations where the incoming and outgoing waves are well defined, and not voltages and currents.

Example 2.3

Let us investigate the *S*-parameters of a microstrip step junction from a 50 Ω line to a 20 Ω line. Again, let us choose a low frequency, say, 1 GHz, where the physical geometry of the discontinuity has little effect, and let us choose line lengths of 5 mm each, with a hypothetical air dielectric used in the microstrip. The reference R is 50 Ω.

Approach 1: Using the ABCD matrix and well-known conversion formulae (Pozar), we get

$$[S] = \begin{bmatrix} 0.11\angle -110.7° & 0.99\angle -14.7° \\ 0.99\angle -14.7° & 0.11\angle -98.7° \end{bmatrix}$$

These values are intuitively satisfying: since the whole circuit is much smaller than a wavelength, it may be regarded as a simple connection from input to output, in which case $S_{11} = S_{22} = 0$ and $S_{12} = S_{21} = 1$. The result is not very different, and we would expect a similar result for a *different R as well*.

Approach 2: Using a standard commercially available simulator (CST – (CST)) and taking a substrate with $\varepsilon_r = 1$ and thickness = 0.5 mm, the result is roughly

$$[S] = \begin{bmatrix} 0.46\angle -153° & 0.88\angle 1° \\ 0.88\angle 1° & 0.46\angle 24° \end{bmatrix}$$

The reason for the difference is that the simulator *does not start from voltages and currents*. Instead, it starts from incident and reflected waves, so evidently some assumption has to be made regarding the transmission lines on which these waves exist. It can be seen from the relevant documentation (CST) that it uses a waveguide port where there is *no discontinuity at the port*. So, the external transmission lines (like the line of characteristic impedance R in Figure 2.6) are obtained by extending the cross section of the port—there is no option to choose a user-defined R (there is the option of "renormalizing," but that is post-simulation). This leads to $R = 50\ \Omega$ at port 1 and $20\ \Omega$ at port 2. Such a situation falls outside our definition, and we have to adopt a more general definition of S-parameters, known as generalized S-parameters. Note that with this new definition, a trivial circuit, consisting of a direct connection of input to output, need not show the intuitively satisfying result: $S_{12} = S_{21} = 1$.

2.1.2 Generalized S-Parameters

Going back to defining equations (1) and (2), the use of $\sqrt{R}$ in defining a_i and b_i may appear to be unusual. In these definitions, it is clear from (2) that $\sqrt{R}$ could have been omitted with no change in the S-parameters. The only reason why it has been kept is to provide continuity with the generalized S-parameters.

Generalized S-parameters are not of much importance in traditional microwave measurements. They are, however, very useful in important areas of microwave engineering like amplifier design and structure simulation, and a brief description is appropriate.

The commonly followed approach to defining generalized S-parameters (following Ha (1981)) again starts by assuming that voltages and currents are well defined at the N ports.

Define a set of N complex numbers with positive real parts: $Z_1, \dots, Z_N$. Let $Z_i = R_i + jX_i$ for $i = 1$ to N.

As before, define the variables (complex) $a_1, \dots, a_N$ and $b_1, \dots, b_N$ by

$$a_i = (V_i + Z_i I_i)/(2\sqrt{R_i}) \quad \text{and} \quad b_i = (V_i - Z_i^* I_i)/(2\sqrt{R_i}) \text{ for } i = 1, \dots, N \quad (2.7)$$

$*$, as usual, denotes complex conjugate.

Then the $N \times N$ S-matrix is defined through the same earlier relation:

$$\begin{bmatrix} b_1 \\ \vdots \\ b_N \end{bmatrix} = \begin{bmatrix} S_{11} & \cdots & S_{1N} \\ \vdots & \ddots & \vdots \\ S_{N1} & \cdots & S_{NN} \end{bmatrix} \begin{bmatrix} a_1 \\ \vdots \\ a_N \end{bmatrix} \quad (2.8)$$

Now it is easily shown that the generalized S-matrix of a direct connection from port 1 to port 2 is indeed close to that obtained earlier by simulation in

Example 2.3, provided $Z_1 = 50$ and $Z_2 = 20$ (imaginary parts = 0). This is left as an exercise for the student.

The terms a_i and b_i are called power waves and in general *do not* admit of any simple physical interpretation. It is tempting to relate them to waves propagating on lossy transmission lines connected to the ports with characteristic impedances $Z_1, \ldots, Z_N$, but due to the Z_i^* term in the definition of b_i, this does not work. Even more importantly, if there is a cascading of components (say, two ports), then b_2 for the first component is *not the same* as a_1 for the second component, again due to the Z_i^* term.

The above definition does obey certain relations concerning power:

1. The power dissipated in the network is

$$\frac{1}{2}\sum_{1}^{N}|a_i|^2 - \frac{1}{2}\sum_{1}^{N}|b_i|^2 \tag{2.9}$$

2. If all b_i are 0, then (from (2.9)) the power dissipated in the network is

$$\frac{1}{2}\sum_{1}^{N}|a_i|^2 \tag{2.10}$$

If we adopted an alternate definition of b_i as $b_i = (V_i - Z_i I_i)/(2\sqrt{R_i})$, we will indeed be able to identify it as a wave on a transmission line, but the above power relations will not be satisfied. In fact, with such a definition, we will end up with certain unusual results, as the next example will show.

Example 2.4

This is an example of voltage, current, and power using generalized *S*-parameters.

Consider the following two cases for what is effectively a one-port (port 2 is terminated in Z_2), as shown in Figure 2.8.

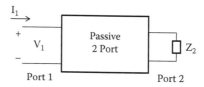

FIGURE 2.8
Voltage, current, and power.

CASE I

Define

$$a_i \triangleq \frac{V_i + Z_i I_i}{2\sqrt{R_i}}, \quad i = 1, 2$$

and

$$b_i \triangleq \frac{V_i - Z_i I_i}{2\sqrt{R_i}}, i = 1, 2$$

Peak, and not root mean square (rms), is used here. Z_1 and Z_2 are two user-defined complex numbers with positive real parts R_1 and R_2.

Let us evaluate

$$D_1 = \frac{1}{2}|a_1|^2 - \frac{1}{2}|b_1|^2$$

$$D_1 = \frac{1}{8R_1}[(V_1 + Z_1 I_1)(V_1 + Z_1 I_1)^* - (V_1 - Z_1 I_1)(V_1 - Z_1 I_1)^*]$$

$$= \frac{2}{8R_1}[V_1(Z_1 I_1)^* + V_1^*(Z_1 I_1)] = \frac{1}{2R_1}\text{Re}[(V_1 I_1^*)Z_1^*]$$

Now power delivered into port $1 = \frac{1}{2}\text{Re}(V_1 I_1^*)$, which is not equal to D_1.

CASE II

Define

$$a_i = \frac{V_i + Z_i I_i}{2\sqrt{R_i}}, \quad i = 1, 2$$

$$b_i = \frac{V_i - Z_i^* I_i}{2\sqrt{R_i}}, \quad i = 1, 2$$

Z_1 and Z_2 stay the same as in case I.

Now D_1 becomes

$$\frac{1}{8R_1}[(V_1 + Z_1 I_1)(V_1 + Z_1 I_1)^* - (V_1 - Z_1^* I_1)(V_1 - Z_1^* I_1)^*]$$

$$= \frac{1}{8R_1}[V_1 Z_1^* I_1^* + V_1^* Z_1 I_1 + V_1 Z_1 I_1^* + V_1^* Z_1^* I_1]$$

$$= \frac{1}{8R_1}[V_1 I_1^*(2\,\text{Re}\,Z_1) + V_1^* I_1(2\,\text{Re}\,Z_1)]$$

$$= \frac{2R_1}{8R_1}[V_1 I_1^* + V_1^* I_1] = \frac{1}{2}\text{Re}(V_1 I_1^*)$$

This is indeed the power absorbed into port 1.

Also, S_{11} in case I is

$$\frac{Z_m - Z_1}{Z_m + Z_1} = \frac{R_m + j\ X_m - (R_1 + j\ X_1)}{R_m + j\ X_m + R_1 + j\ X_1}$$

$$= \frac{(R_m - R_1) + j(X_m - X_1)}{(R_m + R_1) + j(X_m + X_1)}$$

So,

$$|S_{11}|^2 = \frac{(R_m - R_1)^2 + (X_m - X_1)^2}{(R_m + R_1)^2 + (X_m + X_1)^2}$$

Suppose $R_m = 10$, $R_1 = 1$, $X_m = 10$, and $X_1 = -10$. Then $|S_{11}|^2 = \frac{81+400}{121} > 1$, even though these are realistic values for passive circuits.

In case II,

$$|S_{11}|^2 = \left|\frac{Z_m - Z_1^*}{Z_m + Z_1}\right|^2 = \frac{(R_m - R_1)^2 + (X_m + X_1)^2}{(R_m + R_1)^2 + (X_m + X_1)^2}$$

This is always <1 for passive circuits, because R_m and R_1 are positive for all passive circuits.

For further details regarding generalized S-parameters, Ha (1981) and the references mentioned therein may be consulted.

2.2 Spectra of Commonly Encountered Signals

One of the important instruments used in microwave measurements is the spectrum analyzer. The mathematical theory behind the spectra normally encountered is treated in detail in textbooks such as Oppenheim (1983) and Papoulis (1984). Here, we will discuss the following topics from a practical and intuitive way, inevitably sacrificing mathematical rigor:

1. Distribution of power at different frequencies in a power signal
2. Power spectral density for deterministic and random signals
3. Commonly encountered power spectral densities in the case of deterministic signals
4. Commonly encountered power spectral densities for random signals

2.2.1 Distribution of Power at Different Frequencies in a Power Signal

Most signals encountered in this field are properly described as power signals; that is, they contain finite power but infinite energy (the time intervals for which such signals last are usually large enough to be considered infinite from a measurement point of view, although strictly speaking all signals have a beginning and an end). The proper way to describe the distribution of this power in frequency domain is by using the power spectral density.

Let $v(t)$ be a voltage signal feeding a resistor R. We do not assume that $v(t) = 0$ outside an interval $-T_0$ to T_0, so in general, v(t) may not have a Fourier transform in the conventional sense, even though we will assume that $v(t)$ is bounded. The energy dissipated in R is

$$\int_{-\infty}^{\infty} \{v^2(t)/R\}dt$$

which is typically not finite.

However, the power dissipated in R, more precisely, the time-averaged power,

$$p(T) = \frac{1}{T} \int_{-T/2}^{T/2} \{v^2(t)/R\}dt$$

may be (and in practice is always found to be) finite, however large T may be.

If we investigate the conditions when $\lim_{T \to \infty} p(T)$ exists rigorously, it is a difficult problem, which is outside the scope of the present work.

We will adopt a more practical engineering approach toward examining such power relations, which may not always be mathematically rigorous.

Lets us focus on the following problem (Figure 2.9): the band-pass filter is an idealized one:

1. For $f_0 - \frac{B}{2} < |f| < f_0 + \frac{B}{2}$ its output voltage = input voltage.
2. For $|f|$ outside the range $f_0 - \frac{B}{2}$ to $f_0 + \frac{B}{2}$, the output is 0.

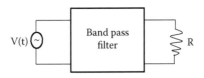

FIGURE 2.9
Interpreting power spectral density.

We would now like to evaluate the power dissipated in R. The problem is not difficult if the Fourier transform of $v(t)$,

$$\int_{-\infty}^{\infty} v(t)\, e^{-j2\pi ft}\, dt$$

existed. Unfortunately, $v(t)$ is not known to be absolutely integrable; i.e.,

$$\lim_{T\to\infty} \int_{-T/2}^{T/2} |v(t)|\, dt$$

is not known to be finite.

So, we start from a truncated version of $v(t)$:

$$v_T(t) \triangleq \begin{cases} v(t) & \text{if } -\dfrac{T}{2} < t < \dfrac{T}{2}, \quad \text{and} \\[2mm] 0 & \text{otherwise} \end{cases} \tag{2.11}$$

$v_T(t)$ is obviously absolutely integrable.

Let

$$V_T(f) = \int_{-\infty}^{\infty} v_T(t) e^{-j2\pi ft}\, dt \tag{2.12}$$

Obviously, we also have

$$V_T(f) = \int_{-T/2}^{T/2} v_T(t) e^{-j2\pi ft}\, dt \tag{2.13}$$

Let the load voltage be $y(t)$. There is no reason to expect $y(t)$ to be 0 outside the interval $\frac{-T}{2}$ to $\frac{T}{2}$ (or any other interval for that matter).

Assuming that $Y(f)$, the Fourier transform of $y(t)$, exists,

$$Y(f) = \begin{cases} V_T(f), & \text{if } f_0 - \dfrac{B}{2} < |f| < f_0 + \dfrac{B}{2} \\[2mm] 0, & \text{otherwise} \end{cases} \tag{2.14}$$

Now

$$y(t) = \int_{-\infty}^{\infty} Y(f)e^{j2\pi ft} \, df$$

$$= \int_{-f_0 - \frac{B}{2}}^{-f_0 + \frac{B}{2}} V_T(f)e^{j2\pi ft} \, df + \int_{f_0 - \frac{B}{2}}^{f_0 + \frac{B}{2}} V_T(f)e^{j2\pi ft} \, df \tag{2.15}$$

The energy dissipated in R is

$$E = \int_{-\infty}^{\infty} \{y^2(t)/R\} \, dt \tag{2.16}$$

Using (2.15), this can be evaluated (Appendix 1) as

$$E = 2B\{|V_T(f_0)|^2/R\}$$

assuming that B is small enough so that $V_T(f)$ is constant in

$$f_0 - \frac{B}{2} \text{ to } f_0 + \frac{B}{2}$$

This prompts the definition of power spectral density of $v(t)$ as

$$S(f_0) \triangleq \left[\lim_{T \to \infty} \left\{ \frac{E}{T} \right\} \right]/(2B)$$

Notice that the bandwidth was actually $2B$, half in positive frequencies and half in negative.

So,

$$S(f_0) = \lim_{T \to \infty} \frac{1}{TR} |V_T(f_0)|^2$$

or in more standard notation:

$$S(f) = \lim_{T \to \infty} \frac{1}{TR} |V_T(f)|^2$$

R is often omitted from the definition (i.e., taken to be 1), but we have retained it, as practical spectrum analyzers usually work with a value $R = 50 \, \Omega$.

Of course, we have assumed that this limit exists.

It is easy to verify that it does not exist in many simple cases, say:

$$v(t) = \cos(2\pi f_0 t)$$

In this case,

$$S(f_0) = \lim_{T \to \infty} \frac{1}{TR} |V_T(f_0)|^2 \to \infty$$

But, even in this case, another related quantity,

$$P(f_0) \triangleq \lim_{T \to \infty} \frac{1}{TR} \int_{f_0 - B/2}^{f_0 + B/2} |V_T(f)|^2 \, df$$

does exist (for arbitrarily small B). This has units of W. To obtain a physically meaningful quantity, we should consider $P(-f_0)$ also, which is equal to $P(f_0)$.

Until now we have not inquired whether $v(t)$ is a deterministic signal (such as $\cos(2\pi f_0 t)$) or random, like noise.

2.2.2 Power Spectral Density for Deterministic and Random Signals

We have examined the concept of power spectral density (PSD) of a deterministic signal in terms of its Fourier transform after truncation.

For a real random voltage signal, an obvious extension is the following definition:

$$S_x(f) = \lim_{T \to \infty} \frac{1}{TR} E\left\{|X_T(f)|^2\right\} \tag{2.17}$$

Here E stands for expectation.

The definition is easier to interpret if expressed in the following way:

$$E|X_T(f)|^2 = E\{|X_T(f) X_T^*(f)|\}$$

$$= E\left\{ \left[\int_{-T/2}^{T/2} x(t) e^{-j2\pi ft} dt\right] \left[\int_{-T/2}^{T/2} x(t') e^{-j2\pi ft'} dt'\right]^* \right\}$$

$$= E \int_{-T/2}^{T/2}\int_{-T/2}^{T/2} x(t)x(t') e^{-j2\pi f(t-t')} dt \, dt'$$

$$= \int_{-T/2}^{T/2}\int_{-T/2}^{T/2} E\{x(t)x(t')\} e^{-j2\pi f(t-t')} dt \, dt'$$

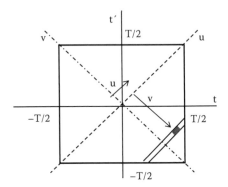

FIGURE 2.10
Relating autocorrelation and power spectral density.

Assuming that $x(t)$ is stationary (actually wide sense stationary (Papoulis (1984))), this becomes

$$I_T(f) = \int_{-T/2}^{T/2} \int_{-T/2}^{T/2} R_x(\tau)e^{-j2\pi f \tau}\, dt\, dt'$$

where $\tau = t - t'$. $R_x(\tau)$ is the autocorrelation of $x(t)$, a real even function. We now use two different independent variables, u and v, to evaluate this integral. These are shown graphically in Figure 2.10.

Consider the small shaded region of size $du \times dv$, located at (u, v).

For this v, the range of u is

$$-\left(\frac{T}{\sqrt{2}} - |v|\right) \quad \text{to} \quad \left(\frac{T}{\sqrt{2}} - |v|\right)$$

Also, looking at the coordinate rotation,

$$u = \frac{t' + t}{\sqrt{2}} \quad \text{and} \quad v = \frac{t' - t}{\sqrt{2}} = -\frac{\tau}{\sqrt{2}}$$

$$\therefore I_T(f) = \int_{v=\frac{-T}{\sqrt{2}}}^{T/\sqrt{2}} \int_{u=-\left(\frac{T}{\sqrt{2}}-|v|\right)}^{\left(\frac{T}{\sqrt{2}}-|v|\right)} R_x(-\sqrt{2}v)e^{j2\pi f \sqrt{2}v}\, du\, dv$$

$$= \int_{v=\frac{-T}{\sqrt{2}}}^{T/\sqrt{2}} R_x(-\sqrt{2}v)e^{j2\pi f \sqrt{2}v}\{\sqrt{2}\, T - 2|v|\}\, dv$$

completing the u-integral.

Now going back to the more usual symbol τ, using the relation $v = -\frac{\tau}{\sqrt{2}}$,

$$I_T(f) = \int_T^{-T} R_x(\tau)e^{-j2\pi f\tau}\{\sqrt{2}\ T - \sqrt{2}\ |\tau|\}\left[-\frac{d\tau}{\sqrt{2}}\right]$$

$$\int_{-T}^{-T} R_x(\tau)e^{-j2\pi f\tau}(T-|\tau|)d\tau$$

For a real-life random voltage signal, $R_x(\tau)$ will be negligibly small for $\tau >$ some τ_{max}, since there is no reason to expect correlation between voltage values observed at widely separated time instants.

Since we will finally take $T \rightarrow \infty$, we can assume that $T \gg \tau_{max}$, so that $T - |\tau| \simeq T$. So,

$$I_T(f) \cong T \int_{-\tau_{max}}^{\tau_{max}} R_x(\tau)e^{-j2\pi f\tau}d\tau$$

Knowing that outside the range $-\tau_{max}$ to τ_{max}, $R_x(\tau) \approx 0$, the limits may as well be changed now to $(-\infty, \infty)$. This gives the Fourier transform of $R_x(\tau)$. Finally, $I_T(f) = T\,P_x(f)$, where $P_x(f)$ is the Fourier transform of $R_x(\tau)$.

Going back to Equation (2.17),

$$S_x(f) = \frac{1}{R}[P_x(f)]$$

$$= \frac{1}{R}\int_{-\infty}^{\infty} R_x(\tau)e^{-j2\pi f\tau}d\tau$$

This is a well-known result; the above derivation is actually a simplified version of the Wiener-Khinchin theorem.

For microwave measurements more general forms of the theorem are not required.

2.2.3 Commonly Encountered Power Spectral Densities in the Case of Deterministic Signals

For deterministic signals, it has been seen earlier that the power spectral density can be arrived at using the Fourier transform of the truncated signal. However, we will arrive at the same results in another way using the autocorrelation function (say, for the voltage signal $v(t)$), defined in the following way:

$$R_v(\tau) = \frac{1}{T} \int_{-T/2}^{T/2} v(t)v(t-\tau)dt \qquad (2.18)$$

Here we will take T to be very large (i.e., in the limit when $T \to \infty$).

As we have seen, the single-sided power spectral density can be calculated from the Fourier transform as

$$S_v(f) = \frac{2}{R} \int_{-\infty}^{\infty} R_v(\tau)e^{-j2\pi ft}d\tau \qquad (2.19)$$

R is the resistor loading the real voltage signal $v(t)$. The final result will be evaluated only in positive f—properties of the Fourier transform ensure that $S_v(f)$ is a real positive even function of f.

Example 2.5

Let $v(t) = A \cos(2\pi f_0 t)$.

To evaluate the autocorrelation, it is convenient (though not mandatory, as can easily be verified) to take $T = NT_0$, where $T_0 = 1/f_0$.

Now,

$$R_v(\tau) = \frac{A^2}{T} \int_{-NT_0/2}^{NT_0/2} \cos(2\pi f_0 t) \cos(2\pi f_0 (t-\tau))dt$$

$$= \frac{A^2}{2NT_0} \int_{-NT_0/2}^{NT_0/2} [\cos(2\pi f_0(2t-\tau)) + \cos(2\pi f_0\tau)]dt$$

$$= \frac{A^2}{2NT_0}(NT_0)\cos(2\pi f_0\tau) = \frac{A^2}{2}\cos(2\pi f_0\tau)$$

which does not change when N becomes large. So,

$$S_v(f) = \frac{A^2}{2R}\delta(f - f_0)$$

gives the *single-sided* power spectral density, if only positive frequencies are to be considered.

This expresses the well-known fact about the above sinusoidal voltage signal: the entire power of $(A^2/2R)$ is concentrated in the single frequency f_0.

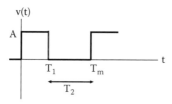

FIGURE 2.11
Periodic train of pulses.

Example 2.6

In this example we demonstrate the train of rectangular pulses.

Let $v(t) = A$ if $nT_m < t < nT_m + T_1$ and $= 0$ if $nT_m + T_1 < t < (n+1)T_m$; n is any integer. This is a periodic train of rectangular pulses, as shown in Figure 2.11.

We would like to evaluate the autocorrelation of $v(t)$, evaluated as

$$\lim_{T \to \infty} \left[\frac{1}{T} \int_{\frac{-T}{2}}^{\frac{T}{2}} v(t)v(t - \tau)dt \right]$$

Notice that $v(t)$ and $v(t-\tau)$ are both periodic functions of t, so their product is also periodic.

Hence, the averaging operation can be carried out over an interval $T = NT_m$, where N will not change the result.

So, we use $N = 1$.

Also, the result is periodic in τ, since $v(t - \tau) = v(t - \tau - N\,T_m)$. So,

$$R_v(\tau) = \frac{1}{T_m} \int_{0}^{T_m} v(t)v(t - \tau)dt$$

CASE I: $T_1 > T_2$

Let us sketch $v(t - \tau)$ for different values of τ, as shown in Figure 2.12. Examining

$$\frac{1}{T_m} \int_{0}^{T_m} v(t)v(t - \tau)dt$$

for these cases, we conclude:

1. $R_v(\tau)$ has a peak of $A^2 \frac{T_1}{T_m}$ at $\tau = 0$.
2. $R_v(\tau)$ decreases linearly with τ up to $\tau = T_2$; $R_v(T_2) = A^2 \frac{T_1 - T_2}{T_m}$.

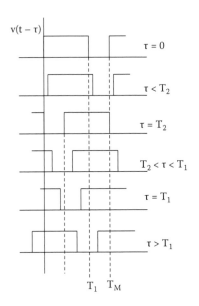

FIGURE 2.12
Shifted versions of $v(t)$: >50% duty cycle.

3. $R_v(\tau)$ stays constant for $T_2 < \tau < T_1$.
4. $R_v(\tau)$ rises linearly with τ, from $A^2 \frac{T_1-T_2}{T_m}$ at $\tau = T_1$ to $A^2 \frac{T_1}{T_m}$ at $\tau = T_m$.

So, we get the autocorrelation shown in Figure 2.13.

CASE II: $T_2 > T_1$

Now the plots of $v(t - \tau)$ are as in Figure 2.14.
   Obviously, this time $R_v(\tau)$ is as shown in Figure 2.15.

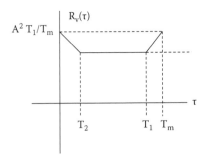

FIGURE 2.13
Autocorrelation of $v(t)$.

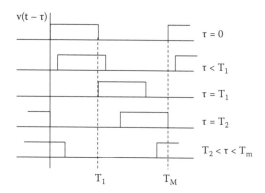

FIGURE 2.14
Shifted versions of $v(t)$: <50% duty cycle.

Before going on to the PSD, it will be desirable to expand these in Fourier series:

$$R_m(\tau) = \sum_{-M}^{M} C_i e^{jif_m\tau}$$

where $f_m = 1/T_m$.

Evaluating C_i in the usual way, for both cases we get:

$$C_0 = (AT_1/T_m)^2$$

And for $i \neq 0$,

$$C_i = (A/i\pi)^2 \, [\sin^2(\pi i \, T_x/T_m)]$$

where T_x is either of T_1 and T_2 (the result is clearly the same).

The number of harmonics M should ideally be infinite, but when considering a modulated signal like $m(t)\cos(2\pi f_0 t)$, M is limited so that $Mf_m \ll f_0$. These keep the expressions simple and also agree with waveforms used in practice.

Also, for $T_1 = T_2$ (50% duty cycle) the even harmonics are all 0.

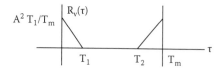

FIGURE 2.15
Autocorrelation of $v(t)$ for <50% duty cycle.

Taking the Fourier transforms, we get the power spectral density of $v(t)$:

$$S_v(f) = \frac{1}{R} \sum_{i=-M}^{M} C_i \delta(f - if_m)$$

There is another way to look at this example: it is easy to expand $v(t)$ in Fourier series:

$$v(t) = \sum_{-M}^{M} v_i e^{jif_m t}$$

where

$$v_i = \frac{A}{\pi i} \sin\left(\frac{\pi i T_1}{T_m}\right) e^{-j\pi i T_1 / T_m}$$

If we *assume* that each term contains power $|v_i|^2 / R$, concentrated at $i * f_m$, expressing this using impulse functions we get the same result for the power spectral density. Of course, the terms in the series are complex in this case other than for $i = 0$, so assigning a power (in watts) to each term requires further justification. This is conveniently done by using Fourier series expansion with sin() and cos() functions, thus considering only positive frequencies and effectively calculating the single-sided PSD. Since the double-sided power spectral density is symmetric about $f = 0$ (even function), the above result can be established.

Relating the PSD of periodic functions to their Fourier series can be done more rigorously by relating the Fourier transform of the truncated signal, the Fourier transform of the actual signal-containing impulses, and the Fourier series coefficients.

The details are the subject of some of the problems.

The double-sided power spectral density of the modulated signal, $p(t) = v(t) \cos(2\pi f_0 t)$, when $mf_m \ll f_0$, can be calculated in the same way, by going through the autocorrelation. The calculations are somewhat tedious, but the result is closely related to the PSD of the unmodulated pulse train:

$$S_p(f) = \frac{1}{2R} \sum_{i=-M}^{M} C_i \delta(f - f_0 - if_m) + \frac{1}{2R} \sum_{i=-M}^{M} C_i \delta(f + f_0 - if_m)$$

2.2.4 Commonly Encountered Power Spectral Densities in the Case of Random Signals

Example 2.7

Let $r(t)$ be a train of rectangular pulses, each of duration T. The height of each pulse is A or $-A$ (V) with equal probability, and the height of any pulse is independent of the height of any other pulse. The load is a resistor R.

The analysis is well known in textbooks and is left as an exercise (problem 4). The PSD is

$$S_r(f) = A^2 T \left[\frac{\sin(\pi f T)}{\pi f T} \right]^2 \frac{1}{R}$$

Example 2.8

In this example, we show amplitude modulation (AM) noise.

Let $x(t) = \{A + n(t)\} \cos 2\pi f_0 t$ be a voltage signal feeding a resistor R. $n(t)$ is a noise signal that is:

1. Zero mean.
2. Slowly varying; i.e., the power spectral density $S_n(f)$ is not negligible only in a small range $-f_n$ to f_n where $f_n \ll f_0$.
3. $S_n(f) = \frac{1}{R} \times$ Fourier transform of $R_n(\tau)$.

To evaluate $R_x(\tau)$ and then $S_x(f)$, we proceed as follows: $R^x(\tau)$ is evaluated by averaging over a long time interval—we are invoking ergodicity (Papoulis 1984).

$$R_x(\tau) = \frac{1}{T} \int_{-T/2}^{T/2} \{A + n(t)\} \cos(2\pi f_0 t) \{A + n(t - \tau)\} \cos(2\pi f_0(t - \tau))\ dt$$

Choose $T = 2NT_0 = \frac{2N}{f_0}$, where N is large. So,

$$R_x(\tau) = \frac{1}{T} \int_{-T/2}^{T/2} \{A^2 + An(t) + An(t - \tau) + n(t)n(t - \tau)\} \cos(2\pi f_0 t) \cos(2\pi f_0(t - \tau)) dt$$

Now we use an approximation. Since $n(t)$ is slowly varying, let us suppose that it is constant in any single time period of the carrier; i.e., $n(t)$ is constant for $t = i\,T_0$ to $(i + 1)T_0$. Call this constant n_i. The integral has four terms, say, R_{x1}, R_{x2}, R_{x3}, and R_{x4}. So,

$$R_{x1}(\tau) = \frac{A^2}{2T} \int_{-T/2}^{T/2} \{\cos(4\pi f_0 t - 2\pi f_0 \tau) + \cos(2\pi f_0 \tau)\} dt$$

$$= \frac{A^2}{2} \cos(2\pi f_0 \tau)$$

$$R_{x2}(\tau) = \frac{A}{T} \sum_{i=-N}^{N-1} \int_{i\,T_0}^{(i+1)T_0} n(t)\left(\frac{1}{2}\right)\{\cos(4\pi f_0 t - 2\pi f_0 \tau) + \cos(2\pi f_0 \tau)\}dt$$

$$= \frac{A}{2T} \sum_{i=-N}^{N-1} \cos(2\pi f_0 \tau)(n_i)(T_0)$$

$$= \frac{A}{2} \cos(2\pi f_0 \tau)\frac{1}{2NT_0} \sum_{i=-N}^{N-1} n_i T_0$$

$$= \frac{A}{2} \cos(2\pi f_0 \tau)\frac{1}{2T} \int_{-T}^{T} n(t)dt$$

$= 0$ since n(t) is supposed to be 0 mean

Similarly, $R_{x3} = 0$.

For R_{x4}, let us say that $\{n(t)\,n(t - \tau)\}$ is roughly constant in one carrier cycle and has a value of P_i for $i\,T_0 < t < (l + 1)\,T_0$. So,

$$R_{x4}(\tau) = \frac{1}{T} \sum_{i=-N}^{N-1} \int_{i\,T_0}^{(i+1)T_0} (P_i)\left(\frac{1}{2}\right)\{\cos(4\pi f_0 t - 2\pi f_0 \tau) + \cos 2\pi f_0 \tau\}dt$$

$$= \frac{1}{2NT_0}(\cos 2\pi f_0 \tau)\sum_{i=-N}^{N-1} \left(\frac{P_i}{2}\right)T_0$$

$$= \frac{1}{2}\cos(2\pi f_0 \tau)\left(\frac{1}{T}\right)\left[\int_{T/2}^{T/2} n(t)n(t - \tau)dt\right]$$

$$= \frac{1}{2}\cos(2\pi f_0 \tau)R_n(\tau)$$

Finally,

$$R_x(\tau) = \frac{A^2}{2}\cos(2\pi f_0 \tau) + \frac{R_n(\tau)}{2}\cos(2\pi f_0 \tau)$$

So,

$$S_x(f) = \frac{A^2}{4R}\left[\delta(f - f_0) + \delta(f + f_0)\right] + \frac{1}{4}\left[S_n(f - f_0) + S_n(f + f_0)\right] \qquad (2.20)$$

which gives the double-sided PSD of x(t)

Example 2.9

This example shows phase noise.

Let $x(t) = A \cos(2\pi f_0 t + \phi(t))$ be a voltage signal feeding a resistor R. Here $\phi(t)$ is noise, with the properties:

1. $\phi(t)$ is zero mean.
2. $\phi(t)$ is small, loosely speaking, $|\phi| \ll 2\pi$.
3. $\phi(t)$ is slowly varying; $S_\phi(f)$ is not negligible in only a small range $-f_\phi$ to f_ϕ, where $f_\phi \ll f_0$.
4. The PSD of ϕ is simply the Fourier transform of $R_\phi(\tau)$. A factor $(1/R)$ is *not* included.

Now, $x(t) = A \cos 2\pi f_0 t \cos\phi(t) - A \sin(2\pi f_0 t) \sin\phi(t)$. Since ϕ is small, this becomes:

$$x(t) \cong A \cos(2\pi f_0 t) - A \phi(t) \sin(2\pi f_0 t)$$

Proceeding as in the earlier case,

$$S_x(f) = \frac{A^2}{4R}[\delta(f - f_0) + \delta(f + f_0)] + \frac{A^2}{4R}[S_\phi(f - f_0) + S_\phi(f + f_0)] \quad (2.21)$$

2.3 Microwave Filters and Directional Couplers

2.3.1 Microwave Filters

Microwave filters are essential components in most measurement setups. Low-pass and band-pass filters are the ones commonly used, while high-pass and band-stop filters are rare in measurement setups.

Detailed information is available about microwave filters in the literature (e.g., Mathaei 1980). Here we will only describe a few commercially available filters, in order to highlight the specifications commonly available in real life.

2.3.1.1 Low-Pass Filter

As the first example, let us consider a low-pass filter from Mini-Circuits, part number BLP-450+.

A photograph is shown in Figure 2.16 and the performance is given in Figure 2.17.

FIGURE 2.16
A low-pass filter with BNC co-axial ports.

The important features are:

1. The 3 dB cutoff frequency is 440 MHz, and below 360 MHz the insertion loss (defined as $-20 \log |S_{21}|$) is below 0.5 dB, and above 565 MHz, the insertion loss is >30 dB. While there is no uniform convention as to how *pass-band* and *stop-band* are defined, there is no doubt that 0–360 MHz belongs to the pass-band, and frequencies above 565 MHz belong to the stop-band.

2. What is not apparent from the information above is the frequency to which the stop-band extends. For this particular component, data are supplied until 1800 MHz and the stop-band definitely extends until there. If there is a requirement that even higher frequencies (say, 5 GHz) have to be stopped, this filter may not be satisfactory. In such a case, another low-pass filter such as Mini-Circuits VLF-1800, with a cutoff frequency of 1800 MHz, may be cascaded. This filter has a stop-band extending until at least 7200 MHz, and the combination will satisfy the requirement.

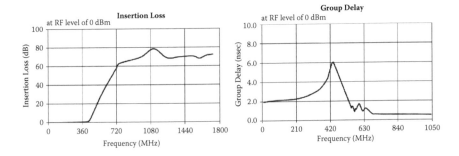

FIGURE 2.17
Low-pass filter performance: (a) insertion loss and (b) group delay.

FIGURE 2.18
A band-pass filter with SMA co-axial ports.

3. The group delay shows that depending on the frequency, the delay from input to output of this filter can range from 2 to 5 ns within the pass-band. A 300-MHz signal has a time period of 3.3 ns, so the group delay variation is substantial. This parameter can be critical in communication networks, but modern instruments usually work using one frequency at a time, so it is easy to compensate for this variation.

2.3.1.2 Band-Pass Filter

For the next example, let us consider the band-pass filter—part number VBF-2555+ from Mini-Circuits. A photograph is shown in Figure 2.18, and the performance (insertion loss) is shown in Figure 2.19.

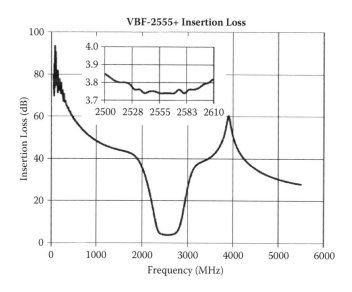

FIGURE 2.19
Insertion loss of the band-pass filter.

Here the pass-band is small, and the group delay is not of much concern. The major issue with this filter is the insertion loss going down from 4 to 5 GHz, and this is likely to give a second pass-band around 6 GHz. This can be corrected by again cascading with a low-pass filter with cutoff around 5 GHz.

Finally, note that the insertion loss here is ~3 dB, which is very high compared to the insertion loss of the low-pass filter. This is a common feature of resonant circuits, and the narrower the fractional bandwidth (higher Q), the more is the loss due to parasitic resistance in the inductors and capacitors used to realize the filter.

2.3.1.3 YIG Tunable Band-Pass Filter

These filters use a special resonator based on a magnetic material (yttrium-iron-garnet [YIG]) whose resonant frequency is tunable over a large range, using a current-carrying coil to apply a tunable dc magnetic field. The resonator commonly takes the shape of a sphere, and the RF signal is coupled to it using printed (microstrip-type) transmission lines in a crossed configuration as shown in Figure 2.20.

This configuration, which is most commonly used, yields a band-pass resonator—the RF signal passes from the input to the output only at the resonant frequency. Other configurations are also possible. Of course, in practice the ball requires some structure to hold it suspended. An actual band-pass filter is shown (top and bottom separated) in Figure 2.21. The tuning coil is clearly visible.

The actual YIG resonators are shown in Figure 2.22, which zooms in on the center of this filter.

It is seen that this filter uses three YIG resonators, which are supported by three miniature ceramic rods oriented diagonally in the figure. The planar transmission lines and the coupling bridges are also clearly visible. Conceptually, the signal flows along the path shown by the arrows at the

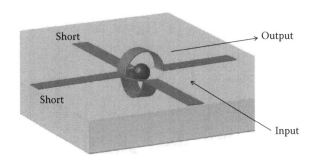

FIGURE 2.20
A YIG resonator with coupling lines.

FIGURE 2.21
A YIG-tuned band-pass filter.

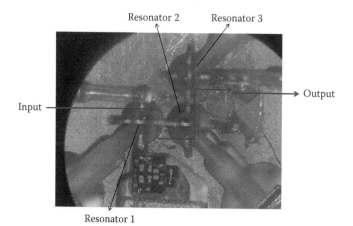

FIGURE 2.22
Practical assembly of YIG resonators.

resonant frequency, which is tunable using the magnetizing coil. This is not quite accurate, as there is also a monolithic microwave integrated circuit (MMIC) switch to select a different path for lower frequencies, but these details are not important here.

Such filters are critical in spectrum analyzers, as will be seen later.

2.3.2 Directional Couplers

As is well known, a directional couple (or just coupler) is a four-port component (passive, and ideally lossless) whose *S*-parameters have the following properties:

1. The diagonal elements are 0.
2. Each column has an additional 0 (other than the diagonal).
3. The magnitude of a nonzero element in any column is either *C* (the coupling) or *T* (the through-port transmission or main line loss).

The number *C* (the coupling) is the main specification of the coupler. Usually it is specified in dB: coupling $= -20 \log_{10} |C|$.

To realize the above properties (specially over a broad frequency range) may be a challenging task, especially if the permissible deviation from ideal values is small.

Ideally, $|C|^2 + |T|^2 = 1$, but in practice there are losses, and $|C|^2 + |T|^2 < 1$. Two other parameters are also used in describing couplers:

1. Isolation: The off-diagonal element in any row that is supposed to be 0 will not be 0 in practice. If its value is some *I*, then $-20 \log_{10} |I|$ is called isolation. Ideally, this should be ∞, but its minimum value in practice (specially over the band of operation) should be significantly higher (in dB) than the coupling.
2. Directivity: This is isolation coupling. In practice this should be at least 10 dB, and most vendors offer products with better than 20 dB directivity.

Usually, the phases of *C* and *T* are not important, except of course in the special case of *C* = *T*. This category is called the hybrid coupler or 3 dB coupler. Further details regarding directional couplers can be found in Pozar (1998). A huge variety of directional couplers are known, and further developments continue to be reported in the literature.

An example is shown in Figure 2.23.

In this case, three of the ports are available (two at the left, and the rightmost) and the fourth port is terminated in a matched load. We may assign the numbers 1 to port IN, 2 to port Through, and 3 to port coupled. Number 4 is assigned to the terminated port, and S_{41} is ideally 0 (actually = *I*). So, as long as we feed the input signal to port 1, no power comes to port 4 and this

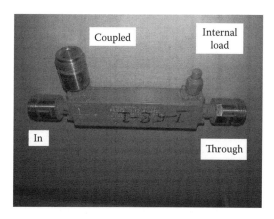

FIGURE 2.23
A directional coupler with co-axial ports.

port is of no use; this is why it is terminated. It is naturally implied that this coupler is to be used with the restrictions:

1. The input should only be at port 1.
2. Ports 2 and 3 should see close to 50 Ω impedances.

For measurement purposes, couplers are mostly (though not always) used under these conditions.

With the above numbering, we get:

$$S_{21} = T, S_{31} = C \text{ and, by symmetry, } S_{32} = S_{41} = I$$

Typical properties of such a coupler are:

1. Frequency range: 1–12.4 GHz.
2. Insertion loss: 1.3 ± 0.5 dB max. This specifies $|T|$ to be *greater* than 0.81.
3. Coupling: 10 ± 0.5 dB (variation with frequency), *referenced to output*. This is a nonstandard way of specifying coupling. Since the output power is 1.3 ± 0.5 dB below the input, in the worst case the true coupling will be $1.8 + 10.5$, or 12.3 dB. This means that the lowest value of $|C|$ over the given frequency range can be 0.24. Note that 10 dB corresponds to $|C| = 0.316$.
4. Coupling tolerance: ± 1 dB. This relates to the variation of the nominal coupling from a particular coupler to another. So, the number 10 dB can actually be 9 or 11 dB. For 11 dB, $|C|$ can be as low as 0.21.
5. Directivity: 12 dB minimum. Since it is possible for coupling to be as low as 13.3 dB, the isolation for this coupler is greater than 25.3 dB (this corresponds to $|I| = 0.054$).

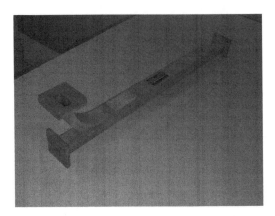

FIGURE 2.24
A waveguide directional coupler.

6. Voltage standing wave ratio (VSWR): 1.45 maximum for ports 1 and 2 and 1.5 for port 3. As we will see in Chapter 3, VSWR at the kth port = $(1 + |S_{kk}|)/(1 - |S_{kk}|)$. Taking the worst case of 1.5, this gives a maximum value of 0.25 for $|S_{kk}|$. So, $|S_{33}|$ should be below 0.25.

A second example of a coupler is shown in Figure 2.24. This has rectangular waveguide ports, and this time all four ports are not available to the user.

2.4 Microwave Mixers, Switches, Attenuators, and Connectors

2.4.1 Microwave Mixers

Microwave mixers are used to effectively multiply two signals. The traditional approach to this is to add the signal and apply the sum to a nonlinear device. The nonlinearity produces many new signals, including the product that is separated with an appropriated filter. Details of mixer theory can be found in Maas (1993).

Today mixing is also achieved using digital techniques (e.g., the exclusive-NOR gate) or the Gilbert cell (Gilbert (1968)), which does not fit well in either digital or analog technology using traditional classification. For the lower frequencies (tens of MHz), actual multiplication can also be done using digital signal processors (DSPs).

In instruments such as network/spectrum analyzers, traditional mixers using Schottky diodes are the most common. These mixers can be classified into three types: single ended, balanced (or singly balanced), and double balanced.

The operation of the three types can be understood by using the following simplifications: let the two inputs be sinusoidal signals:

$V_L \cos(\Omega_L t)$: the larger signal, usually called local oscillator (LO)

$V_R \cos(\Omega_R t)$: the smaller signal, usually called RF

Let us suppose that the nonlinear device generates signals (call these the output $y(t)$) from the sum of the above two signals (call this $x(t)$) through a simple quadratic polynomial relation:

$$y(t) = A + Bx(t) + Cx^2(t)$$

The constant A produces only a DC shift and may be ignored.

Because of the second term, we get the input frequencies Ω_L and Ω_R. Because of the last term, we get the frequencies $2\Omega_L$, $2\Omega_R$, $\Omega_L + \Omega_R$, and $\Omega_L - \Omega_R$. Assuming that Ω_L and Ω_R are close (this is the most commonly encountered case of the downconverter), the difference frequency is much smaller than all other frequencies, and can be separated with a low-pass filter. This is the operating principle of a single-ended mixer. In mixer terminology, the conversion gain (CG) is defined as

$$CG = 20 \log_{10} [(\text{IF power delivered to the load})/(\text{RF power available from source})]$$

The intermediate frequency (IF) is most commonly (but not always) the difference frequency. Diode-based passive mixers usually have CG ~ –10 dB. The noise figure is another important characteristic of a mixer—this will be treated in a later chapter.

A single-ended mixer is rarely used in practice, because the demand on the low-pass filter will be excessive unless the difference frequency is extremely low. As a typical example, V_L may be 1 V, V_R may be 0.1 mV, f_L may be 1 GHz, and f_R may be 1.01 GHz. In this case, the difference frequency signal strength will be the RF signal strength + CG, while the signal at f_L would have a strength somewhat below the LO. So, the filter should ensure that the stop-band attenuation at 1 GHz is more than 80 dB compared to the pass-band at 10 MHz. This will involve a nontrivial effort in filter design, which is not actually necessary.

The problem in the single-ended mixer can be remedied by using a balanced mixer. In its simplest form, the sum and difference of the two inputs are fed to two identical nonlinear devices, and the difference between the two sets of generated signals is taken as the final output. By virtue of the relation

$$(V_L + V_R)^2 - (V_L - V_R)^2 = 4V_L V_R$$

only the sum and difference frequencies are generated at the output due to the square term. Not only is the filtering requirement largely reduced, but it can also be shown that if there is a small noise added to the LO, close to the LO frequency, then the downconversion of this noise is also eliminated. This is important in practice because random small fluctuations in the LO amplitude (called LO AM noise) are inevitable, and lead to sidebands of the LO, which can actually be stronger than the RF.

The balanced mixer operates quite satisfactorily for many applications and is widely used. The realization of the sum and difference signals is achieved using 3 dB couplers (Pozar 1998) or innovative structures such as in the crossbar mixer (Maas).

Because of the linear term in the device input-output relationship, the RF signal will still leak to the output and has to be filtered out. This is not a big problem, since the difference frequency signal is proportional to the RF signal and is not much smaller in a properly designed mixer. Still, it is possible to eliminate this also, by operating one balanced mixer as described above, a second identical one with both LO and RF reversed, and finally adding the two outputs. This is (loosely speaking) the idea behind a double-balanced mixer. The student should work out the mathematical details involved.

Actual double-balanced mixer circuits (Maas 1993) use baluns or transformers and are often used at the lower frequencies. At the higher frequencies (particularly millimeter-wave frequencies), the balanced mixers are standard.

Irrespective of the type, for measurement applications mixers work under the following assumptions:

1. The LO signal is large enough to switch the diodes on and off from one half cycle to the next. It is implied that the switching time of the diodes is small enough. Schottky diodes are invariably used.

2. The RF signal is much smaller than the LO, and the output (IF) is linearly related to the RF. It is often convenient to view the mixer as a linear (though time-varying) component with RF as input and IF as output—its function is to translate the frequency of the RF by an amount equal to the LO frequency.

2.4.2 Microwave Switches

The standard microwave switch is explained functionally in Figure 2.25. This has one RF input and k digital (as good as dc compared to the RF frequency) control lines. Depending on the k-bit 1-0 sequence on the control lines, the input is routed to one of the 2^k outputs. Circuits with $k = 1$ (single-pole double-throw (SPDT)), $k = 2$ (single-pole four-throw (SP4T)), and $k = 3$ (single-pole eight-throw (SP8T)) are commonly used switches.

A slightly different case is the single-pole single-throw (SPST) switch, which has one input, one control line, and one output; here, when the signal

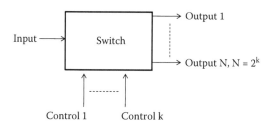

FIGURE 2.25
Function of a microwave switch.

is not routed to the output, it is either absorbed in the switch (absorptive type) or reflected back at the input (reflective type). These behaviors (absorptive/ reflective) apply to the other types of switches as well. For example, if an SPDT switch has routed the input to output 1, then looking into output 2, we may see a high reflection (reflective) or very little reflection (absorptive).

Very broadband (DC 20 GHz) switches are available from manufacturers like Hittite Microwave Corporation, and are widely used in microwave systems for signal routing.

Switching circuits can be realized using p-i-n diodes or FETs. The field-effect transistors (FET)-based switches are more common today. A simplified circuit for an SPDT switch is shown in Figure 2.26. This uses two FETs, only one of which is ON at any instant, due to the inverter.

2.4.3 Microwave Attenuators

The attenuator is perhaps the simplest microwave component. It is a two-port component and is simply described by the S-parameters: $S_{11} = S_{22} = 0$; $S_{21} = S_{12} = A$.

A is a real positive number less than 1, and $(-20 \log_{10} A)$ is the attenuation factor. For a 6 dB attenuator, $A = 0.5$. For a 20 dB attenuator $A = 0.1$.

An attenuator with fixed A can be easily realized with a three-resistor T or pi network.

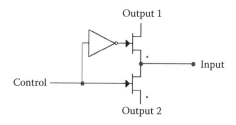

FIGURE 2.26
A microwave SPDT switch using two FETs and an inverter.

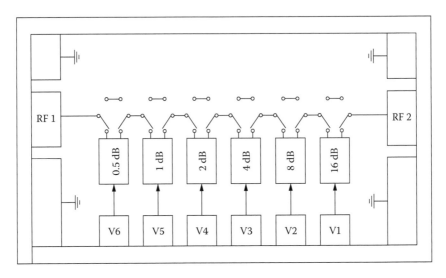

FIGURE 2.27
Conceptual schematic of a digitally variable attenuator.

Variable attenuators require that *A* be tunable, and should cover a wide range of values in response to either an analog control signal or (more commonly) a set of digital control signals.

Keeping *A* fixed (for a particular control signal) over a wide frequency range and also ensuring adequate power handling capability are the challenges faced in the development of attenuators.

Today, attenuators are mostly realized using FETs. Figure 2.27 shows the basic architecture of a 6-bit stepped attenuator which can give up to 31.5 dB attenuation in steps of 0.5 dB. As can be seen, it uses six resistive attenuators with the given attenuation factors and 12 (2 for each resistive attenuator) SPDT switches. Input is at *RF1* and output at *RF2*.

2.4.4 Microwave Connectors and Adaptors

As the final section in this chapter, the most commonly used microwave co-axial connectors are described—low-frequency connectors such as the BNC are not discussed here. Much more detailed information is available from the websites of manufacturers such as Amphenol and Anritsu.

The most commonly used microwave co-axial connector is the SMA. Pictures of this type are shown in Figure 2.28. This is usable until 18 GHz. The diameter of the Teflon dielectric is around 4.5 mm. To understand the terms *inner conductor* and *outer conductor,* recall that the electric field is present between the

(a)   (b)

FIGURE 2.28
(a) SMA male connector. (b) SMA female connector.

inner and outer conductors in a co-axial cable. The inner conductor is of course easy to identify in all the connectors described here.

The SMA connector is compatible with a slightly better connector called the 3.5 mm series (not to be confused with an audio frequency connector also called 3.5 mm connector). Here the dielectric is air and the diameter of the outer co-axial conductor is 3.5 mm. This is shown in Figure 2.29. This type is usable until 26.5 GHz.

The SMA and 3.5 mm types are roughly compatible with the 2.9 mm or *K* connector. However, repeated connections between these types may damage the *K* connector, and this is not recommended. A picture is shown in Figure 2.30. Here the outer conductor diameter is 2.9 mm. This works spurious mode free until 40 GHz.

The *N*-type connector is another type commonly used for somewhat lower frequencies, which can carry higher powers. A picture is shown in Figure 2.31, and an adaptor from *N* to SMA is shown in Figure 2.32. As can be seen, the *N*-male type has an extra outer shell in addition to the outer co-axial conductor. The *N*-type connector may operate up to

FIGURE 2.29
A 3.5 mm co-axial adaptor.

FIGURE 2.30
A female K connector.

18 GHz, but most versions sold commercially work up to 12 GHz or even lower.

Connectors that work at higher frequencies are the 2.4 mm or *Ka* type, which works up to 50 GHz; the 1.85 mm or *V* type, which works up to 67 GHz; and the 1 mm type, which works up to 110 GHz. The first two are mutually compatible, but not with any of the other types. The 1 mm connector is not compatible with any other type. Such numbers (2.4, 1.85, and 1) refer to the *inner diameter of the outer conductor* of the co-axial cylinders between which the field exists (not to be confused with other supporting metallic structures). The V and 1 mm connectors are shown in Figures 2.32 and 2.33 (actually, these are views of an adaptor between these types; the connector geometry is obviously the same whether used in an adaptor or at the end of a cable).

Adaptors (such as the one in Figures 2.32 to 2.34) are available between any two of the co-axial connectors mentioned above.

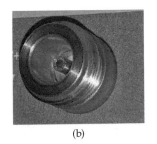

(a)   (b)

FIGURE 2.31
(a) *N*-type male connector. (b) *N*-type female connector.

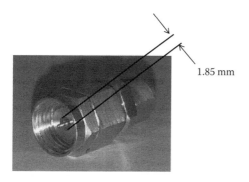

1.85 mm

FIGURE 2.32
A male 1.85 mm (or *V*-type) connector.

FIGURE 2.33
A female 1 mm connector.

FIGURE 2.34
An adaptor from *N*-male to SMA female showing both ends.

2.5 Conclusion

We have presented a simplified description of certain mathematical concepts on which microwave measurements are based. Some of these topics (those related to power spectral density) are normally treated in a different way that is more suitable for further development of them. Here the focus is different, and we believe that the presented approach is more suitable in this context.

The chapter concluded with a description of some practical components that are frequently used in microwave measurements.

Problems

1. Derive the *S*- and ABCD parameters of an ideal transformer with turns ratio *N*:1.

2. What is the Thevenin equivalent of a 1 V source, in series with a 50 Ω resistor, followed by a transmission line of length $\lambda/6$ and characteristic impedances (a) 50 Ω and (b) 10 Ω?

3. A (deterministic) pulse train has 50% duty cycle, with levels 1 and –1 and time period T_0. Derive its power spectral density through the Fourier transform of a truncated version of this signal (avoiding the autocorrelation).

4. Let $r(t)$ be a train of rectangular pulses, each of duration *T*. The height of each pulse is *A* or –*A* (V) with equal probability, and the height of any pulse is independent of the height of any other pulse. The load is a resistor *R*. Derive the PSD, $S_r(f)$.

5. Design a 10 dB attenuator using a resistive *T*, using shunt resistor = R_1 and series resistors = R_2. Reference is 50 Ω.

References

Computer Simulation Technology (CST). www.cst.de.

Gilbert, B. A Precise Four-Quadrant Multiplier with Subnanosecond Response, *IEEE Journal of Solid-State Circuits*, 3(4), 365–373, 1968.

Ha, T.T. *Solid-State Microwave Amplifier Design*. John Wiley & Sons, New York.

Itoh, T. *Numerical Techniques for Microwave and Millimeter-Wave Passive Structures*. Wiley-Interscience, New York.

Maas, S.A. *Microwave Mixers*. Artech House, Norwood, MA, 1993.

Mathaei, G.L., L. Young, and E.M.T. Jones. *Microwave Filters, Impedance-Matching Networks and Coupling Structures*, Artech House, Norwood, MA, 1980.

Papoulis, A. *Probability, Random Variables and Stochastic Processes*. McGraw-Hill, New York, 1984.

Pozar, D.M. *Microwave Engineering*, John Wiley & Sons, New York, 1998.

Oppenheim, A.V., A.S. Willsky, and I.T. Young. *Signals and Systems*. Prentice Hall, New Jersey, 1983.

3

Traditional Measurement Techniques

In this chapter we will study some older measurement techniques, which were used before computers and analog-to-digital converter systems became widespread. These techniques, however, give valuable insight into micro-wave circuits.

3.1 The Power Meter

Measuring the strength of a microwave signal is the starting point for micro-wave measurements (generation of signal is of course equally fundamen-tal—this will be described in a later chapter). A more precise description of strength of a signal is power available from a source. A source here signifies a voltage source in series with some impedance. This, by Thevenin's theo-rem, can be the equivalent circuit representing a much more complex circuit, possibly containing multiple sources.

Recall that power available from a source (with a finite source impedance) is defined as the maximum power that can be dissipated in an appropriate load connected to the source. The maximum power transfer theorem tells us that "appropriate" here means conjugate matched. For such a matched load, half of the power drawn from the voltage source is dissipated in the load, while an equal power is dissipated in the series impedance. So, there is a slight ambiguity when power measurements are made—is it the available power that is being measured, or is it the power absorbed in the instrument (the input impedance of the instrument represents the load that is supposed to be conjugate matched)? Usually measuring instruments are designed to have an input impedance of a standard value (50 Ω is most common), and it is assumed that the source impedance is also of this same value, but these are never very accurate. Moreover, the instruments are normally calibrated to display the power absorbed in them, so the available power is the same as the displayed value only if there is a true impedance match. In case the source impedance is different (but known), simple circuit calculations (left as an exercise for the student) will give the available power, when the power absorbed in a meter with ~50 Ω input impedance is known.

The component used for this measurement is a power detector, which essentially converts a microwave signal to a more easily measured quantity—usually a constant voltage. There are three types of such detectors that are in common use even today: the thermocouple-based, thermistor-based, and diode-based types. Any of these, along with suitable electronics to display the result in a usable manner, forms a power meter. The thermocouple and thermistor sensors cover a range from approximately 1 µW to almost 1 W, while the diode sensors cover roughly from 0.1 nW to around 100 mW. Higher ranges are available using attenuators, while amplifiers (and heterodyning) can sometimes be used for measuring even lower values.

Traditionally the sensor has been assembled as one unit, while the rest of the circuits, the display, and peripheral components (such as power supply) have been assembled as another unit. These are connected by a cable incorporating several wires. Modern power meters usually omit the second unit, and all the required electronics are integrated with the sensor. For display, a computer is used, connected to the sensor through USB.

Calorimeter-based detectors have also been used (Agilent 2006), but they are encountered infrequently today, and will not be described here.

Even though the spectrum analyzer (Chapter 5) is in many ways a superior instrument, there are many cases in which a power meter is mandatory (e.g., ETSI EN 300 328 V1.8.1).

3.1.1 Thermocouple-Based Power Detector

As is well known (Seebeck effect), if two different conductors form a junction that is heated while the other ends of the two conductors are kept cold, a potential difference develops between the two cold ends. The potential difference is roughly proportional to the temperature difference between the hot junction and the cold ends. For practical use, at least one of the conductors should *not* have a very good conductivity (i.e., a copper + aluminum thermocouple is not preferred). This is because the whole effect depends on the generation of additional free carriers (electrons) due to heat, and in metals with a very large free carrier concentration to start with, heat does not alter the number substantially. Additionally, if the heat is generated by dissipating a microwave signal in a resistance located at the junction, the rise in temperature (and hence the potential difference) is proportional to the power of the signal. This is the basis for the thermocouple-based power detector.

For making a practical device starting from this concept, the following points have to be addressed:

1. The microwave signal to be dissipated as heat should see a matched load (conventionally 50 Ω) throughout the band of operation, which should be very wide.
2. The generated dc potential should be isolated from the microwave signal.
3. Circuits should be implemented to measure the dc potential accurately, even when it is extremely small.

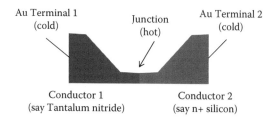

FIGURE 3.1
Approximate layout of the basic thermocouple.

We will see how these issues are addressed in the Agilent 8481A power sensor. The basic thermocouple element itself is realized in the following way in this particular sensor.

Here, the two conductors used are tantalum nitride (TaN, $\rho \sim 10^{-3}$ Ω-cm) and heavily doped ($n+$) silicon with $\rho \sim 10^{-2}$ Ω-cm. Gold contacts are used for external connections (gold has $\rho = 2.4 \times 10^{-6}$ Ω-cm).

The approximate shape of the thermocouple, realized in planar form, is shown in Figure 3.1.

The two conductors are actually formed in two different layers of the planar structure, and there is a thin layer of SiO_2 between them, as shown in Figure 3.2. Contact is achieved through an opening in the SiO_2 layer.

To complete the structure and arrive at something that can be easily fabricated, a few more features are added, as shown in Figure 3.3.

The components here, approximately in the sequence in which they are fabricated (bottom to top), are:

1. One bulk p-silicon frame for providing physical support
2. A thin membrane of n-silicon, in which a particular region is heavily doped to form conductor 2
3. A SiO_2 layer with openings at either end of the $n+$ silicon region
4. TaN film, which makes contact with the $n+$ silicon through one opening
5. Gold contact, which reaches the $n+$ silicon through the other opening
6. Gold contact for the TaN film°

The layers are shown in 3D in Figure 3.4.

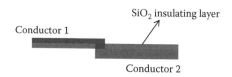

FIGURE 3.2
Cross section of the basic thermocouple.

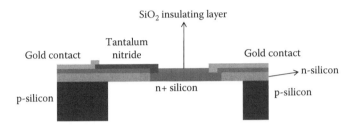

FIGURE 3.3
Cross section of the complete thermocouple structure.

The microwave signal reaches the junction through the gold contacts, and since the contact area is very narrow (n+ silicon and TaN) compared to the rest of the conductors (gold and n+ silicon, gold and TaN), most of the heat is dissipated across n+ silicon and the TaN junction. A photograph of the chip with two thermocouples is shown in Figure 3.5.

The exact shape of the conductors (mainly the TaN; due to small film thickness, this actually contributes most to the resistance, and not the n+ silicon) is tailored to achieve a resistance of 100 Ω between the gold contacts, and two such thermocouples on a single silicon chip are used in a circuit (to be described next) so that they effectively appear in parallel to the microwave input.

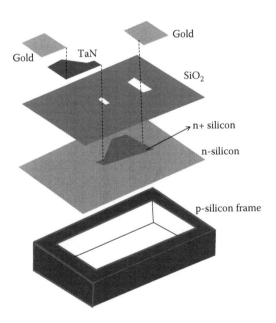

FIGURE 3.4
The different layers in the complete thermocouple structure.

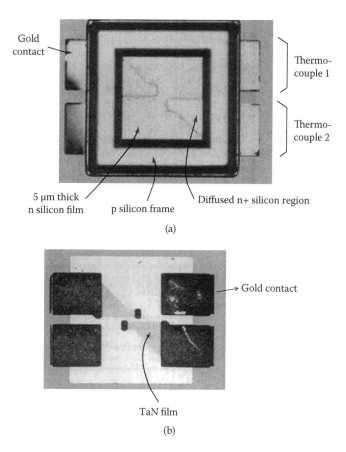

Gold
contact

Thermo-
couple 1

Thermo-
couple 2

5 μm thick
n silicon film   p silicon frame   Diffused n+ silicon region

(a)

Gold contact

TaN film

(b)

FIGURE 3.5

The actual thermocouple chip. (a) Top view. (b) Bottom view. (c) Agilent Technologies 2006. Reproduced with permission.

The two thermocouples on a single chip are used in a circuit shown in Figure 3.6.

The high-value blocking capacitors (Cb) ensure that to the microwave signal, the thermocouples appear in parallel. The dc voltages across the thermocouples add, being in series, and appear at the output. An advantage of this particular configuration is that no inductor is required for radio frequency (RF)–dc isolation. Isolation is achieved purely through the blocking capacitors, and moreover a common ground can be used for both RF and dc. The actual construction is shown in Figure 3.7 (the output capacitor is described as "bypass capacitor").

The final challenge is amplifying the dc output, which can be extremely small (sub-μV) for low microwave powers. This is achieved using a chopper; the complete block diagram is shown in Figure 3.8.

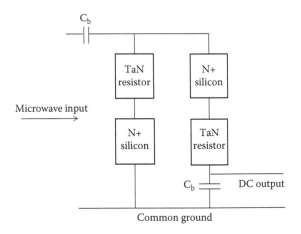

FIGURE 3.6
Circuit using the thermocouples.

The chopper (a simplified circuit is shown in Figure 3.9) effectively connects the input to the output when the trigger is high and gives 0 output when the trigger is low.

It is quite possible that there may be nonidealities in the circuit, and what we get at the output when we expect 0 is not quite 0, and when we expect

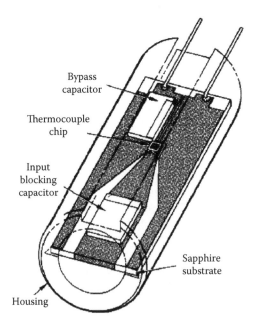

FIGURE 3.7
Construction of the sensor. (c) Agilent Technologies 2006. Reproduced with permission.

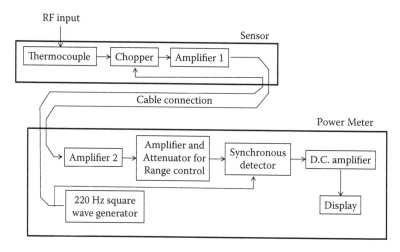

FIGURE 3.8
Building blocks of a thermocouple-based power meter.

the input voltage at the output, again, it may not be quite the input voltage. But the difference between the two levels will be proportional to the input signal, which is what we require. The output of the chopper can be amplified by ac-coupled amplifiers (the amplifiers before the detector are of this type), and since there is no question of dc drift now, these amplifiers can have ample gain, in order to boost the weak dc signal to a desired level. The

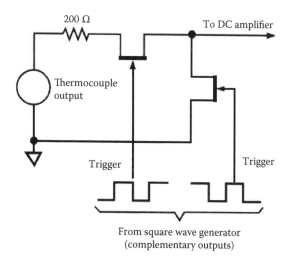

FIGURE 3.9
A simple chopper using FETs.

first amplifier is incorporated with the sensor itself, to avoid degradation of the signal-to-noise ratio (SNR) in the cable (which may be >10 ft in length).

The synchronous detector reverses the operation of the chopper. The same complementary triggers shown in Figure 3.9 are used here, but now these alternately connect the input *or its negative* to the output. This gives back a dc that is equal to the amplitude of the input ac (square wave). If emitter-coupled bipolar junction transistor (BJT) pairs are used in the amplifiers, then output is naturally available in complementary form, which is particularly suitable for such synchronous rectification. The dc signal can be further amplified if required before the display. The display will of course be calibrated suitably to display the power absorbed by the sensor.

The chopper has a limitation from spikes generated at the switching edges. The average value of these spikes is typically nonzero and will effectively add an offset to the final displayed value. In fact, this offset is proportional to the frequency of the trigger pulses (doubling the frequency will give double the number of spikes per second). This offset is taken care of in two ways: (1) the frequency is kept to a low value of 220 Hz, and (2) a "zeroing" facility is added (not shown above). For zeroing, an offset (positive or negative) is added to the thermocouple output, which is manually adjusted with no input connected, until the display reads 0. Fortunately, the offset due to spikes is not dependent on the thermocouple output, so this zeroing procedure nullifies it. The spikes are, however, susceptible to drift, and zeroing may have to be carried out at regular intervals during measurement.

Finally, the power meter also incorporates a 50 MHz oscillator (not shown above) with a precisely controlled amplitude, which is used as a reference for calibrating the sensor. This inherently assumes that the system is linear (i.e., doubling the input power doubles the displayed dc output). This is true to a high degree of accuracy, although the linearity is not as good as the thermistor-based units—details may be found in Agilent (2006).

3.1.2 Thermistor-Based Power Detector

The core element used for this is a thermistor—a resistor composed of a special material (usually based on certain metal oxides) that exhibits a large change of resistance with temperature (4% change per °C is a typical value). For power-detector applications the thermistors used have negative temperature coefficients (the resistance decreases as temperature increases).

The specific instrument that will be described here is the Agilent 432A power meter using the Agilent 478A sensor.

The sensor (also called thermistor mount) contains the simple circuit shown in Figure 3.10. All four resistors shown here are identical thermistors. The capacitors are high-value dc blocks.

Each of the four resistors in Figure 3.10 is actually a thermistor with a value somewhat higher than 100 Ω at room temperature. The actual value will be

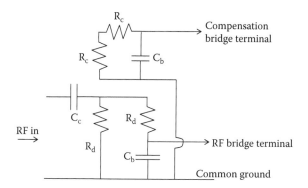

FIGURE 3.10
Circuit of the thermistor sensor.

automatically controlled to exactly 100 Ω using a special circuit shown in Figure 3.11.

In the circuit, two high-gain differential amplifiers (effectively op-amps) are used. Consider the one connected to the RF bridge terminal of the sensor (Figure 3.10).

Noting that the gain is very high, and also that there is negative feedback (from V_r to the − terminal), it is clear (basic op-amp theory—check, for example, Millman and Grabel, 1987) that the two inputs to the op-amp are almost at the same potential. This gives us $[V_r/2] = V_r [R_r/(200 + R_r)]$, where R_r is the resistance between the RF bridge terminal and ground.

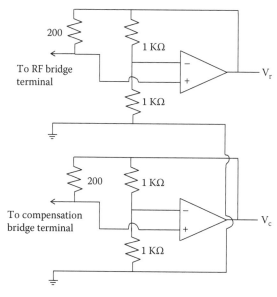

FIGURE 3.11
Driving circuit for thermistor mount.

There are two possibilities: either $V_r = 0$ or $R_r = 200$. The first possibility can be excluded, because the bias arrangements inside the op-amp are such that with both inputs at 0, a moderate positive output is obtained. The conclusion is that Rr = 200. Similarly, Rc = 200, being the resistance between the compensation bridge terminal and ground. Very importantly, *these values are enforced even if ambient temperature changes or RF power is dissipated* in the thermistors (R_d in Figure 3.10).

If the RF power input rises, tending to dissipate more heat, subsequently lowering R_d, the circuit compensates by reducing V_r. This will reduce the dc power dissipation in the two thermistors (R_d). This function is called dc power substitution—the amount of RF power dissipated is equal to the dc power reduction. It maintains the thermistors at the same temperature (and resistance) in spite of RF power variation, and gives the superior linearity of these sensors.

The dc power substitution can be written mathematically as

$$P_{RF} = \frac{Vr0^2}{4R} - \frac{Vr^2}{4R}$$

where V_{r0} is the value of V_r when no RF power is applied. $V_r^2/(4R)$ is the dc power dissipated in the RF bridge ($R = 200\ \Omega$, the dc resistance).

There is an auto-zeroing circuit (not shown) though which an additional offset can be manually incorporated in the op-amp connected to the compensation bridge in order to manually equalize V_r and V_c when there is no RF power input. So, $V_c = V_{r0}$. Finally,

$$P_{RF} = \frac{Vc^2}{4R} - \frac{Vr^2}{4R}$$

This can be easily implemented using modern digital technology. The Agilent 432A power meter uses analog circuits to evaluate $[V_c^2 - V_r^2]$ through the following steps:

1. The sum and difference of V_c and V_r, $(V_c + V_r)$ and $(V_c - V_r)$ are obtained through the circuit in Figure 3.12. The sum is in the form of a dc voltage, but the difference is in the form of a 5 KHz square wave, the difference of high and low levels, which gives $(V_c - V_r)$.

2. The sum signal generated in Figure 3.12 is passed through a pulse-width modulator (PWM) (Millman and Grabel, 1987), i.e., the same 5 KHz square wave in Figure 3.12 has its duty cycle made proportional to the sum signal. Call this waveform the PWM signal.

3. A chopper circuit similar to the one in Figure 3.9 is given as input the difference waveform while its trigger is the PWM signal. The output amplitude of the chopper will be proportional to $(V_c - V_r)$, while the duty cycle of the chopper output will be proportional to $(V_c + V_r)$.

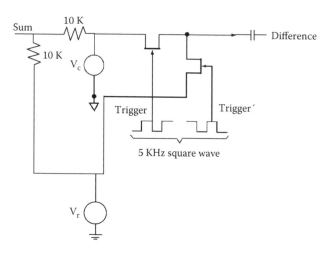

FIGURE 3.12
Sum and difference generator.

4. The chopper output is low-pass filtered. This gives its average value, which is proportional to $(V_c - V_r)(V_c + V_r) = V_c^2 - V_r^2$.

5. A proportionality constant is incorporated by a onetime calibration during manufacture (or subsequent recalibration if required).

It should be kept in mind that these circuits were developed before digital technology and microprocessors became widespread and do not contribute significantly to microwave technology. The display will of course be calibrated suitably to display the power absorbed by the sensor.

3.1.3 Diode-Based Power Sensor

These sensors use Schottky diodes in a circuit that is identical to the half-wave rectifier in configuration, as seen in Figure 3.13.

The Schottky diode has a very complex equivalent circuit at microwave frequencies, and analysis of this circuit is difficult. It can of course be easily simulated, but before that we can arrive at some useful conclusions using simple reasoning.

Suppose we wish to plot the output voltage (it can only be dc since the capacitor C is assumed large enough to short any RF signals) as a function of A (the RF amplitude). We can conclude:

1. The plot passes through the origin—obvious, as with 0 input we will not get any output voltage.

2. The plot gives a smooth curve, since all the equations governing this circuit are known to give smooth (continuously differentiable) functions as solutions.

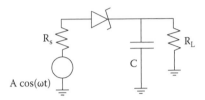

FIGURE 3.13
A Schottky diode-based power detector.

3. The output voltage as a function of A is an even function. To justify this, note that changing the sign of A is the same as delaying the input by a half-cycle, which has no effect on the output that is constant with time.

4. The output rises monotonically as $|A|$ increases—something that is intuitively expected from diode behavior.

All the above are, of course, easily verified by simulation (any SPICE-type simulator). Basic mathematics tells us that such a curve is represented close to the origin by the equation $V_L = CA^2$.

Since the power available from the voltage source along with R_s is $[A^2/(8\,R_s)]$, we conclude that the output voltage is proportional to the power available, and indeed this functions as a practical power sensor. The display associated with the sensor will of course be calibrated to display the available power, in order to stick to our earlier convention of displaying the power absorbed. It is also clear that such operation will be restricted to low input powers. To estimate how low the power should be, we carry out a simple simulation on the circuit shown in Figure 3.14.

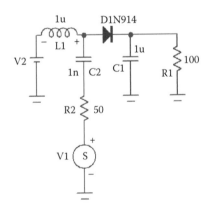

FIGURE 3.14
Simple diode power detector circuit.

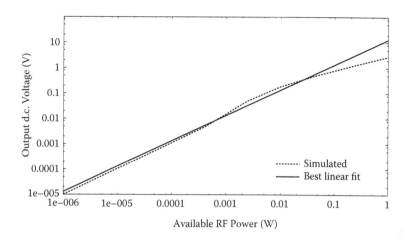

FIGURE 3.15
Simulated behavior of the diode power detector.

A 1 nF capacitor blocks the dc bias from reaching the RF source, and a 1 μH inductor prevents RF signals from reaching the battery, which is used to bias the diode with a low current of 2 mA. The RF frequency is 1 GHz. Note that this particular diode (1N914) is *not* a Schottky diode—it was used since the SPICE model was readily available. In any case, the qualitative features of this type of power sensor are brought out.

The simulated plot of output voltage (across R_1) as a function of available RF power is shown in Figure 3.15. We can see that up to an available power of 40 mW, we do indeed have a linear relationship between RF power available and dc power output. Beyond this, however, the output voltage increases in a sublinear fashion as the available power increases. With modern technology, the measured diode characteristics can be stored with the sensor in a read-only memory (ROM) as a lookup table, and the power for any output dc voltage can be looked up in the ROM, including for higher powers where linearity is not maintained. The upper limit is then imposed only by damage to the diode.

The output voltage for 0 RF power is of course not 0 in this case because there is a dc bias that is applied to the diode. If this value is V_o, then V_o has to be subtracted from all readings.

Low-barrier Schottky diodes (Szente et al., 1976) and planar doped barrier (PDB) Schottky diodes (Malik et al., 1980) achieve far superior performance even without biasing. For example, using PDB diodes, an available RF power of 0.1 μW will give 50 μV dc output. Of course, there is a limitation on the higher powers; linearity (power to voltage) is lost well before 1 mW available RF power. Additionally, the very low voltages are susceptible to corruption due to thermoelectric and other effects. Consequently, the modified circuit shown in Figure 3.16 forms the basis for power sensors like Agilent E4412.

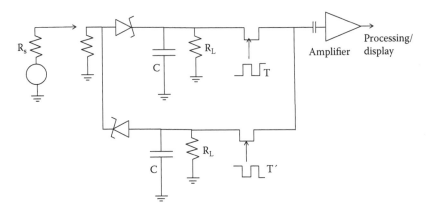

FIGURE 3.16
Balanced power detector.

This uses the same concept as in Figure 3.12: the outputs from the diodes are ideally complementary (negatives of each other) since the reasoning used to explain the operation of Figure 3.13 works equally well with the diode reversed, only there is a change of sign in the output. However, the nonidealities (like thermoelectric effect) lead to offsets that add equally to both outputs, and hence get canceled when their difference is amplified through an ac-coupled amplifier as shown in Figure 3.16.

To extend the operation to higher powers, stacked diodes are used as shown in Figure 3.17. In this configuration, four times the power can be applied, since the allowed voltage across a diode pair is twice that for a single diode. Of course, sensitivity at low powers is lost, so this configuration has to be used in conjunction with the single-diode sensor to achieve both high sensitivity at low powers and usability at high powers.

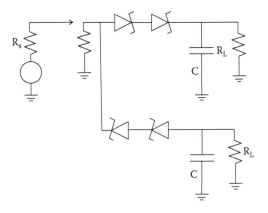

FIGURE 3.17
Stacking diodes to operate at higher powers.

The resistor preceding the diodes improved the matching of the input; especially at low powers, the diodes show a high effective impedance to the RF input. To reduce reflection of the input signal, this resistance offers a certain degree of matching—there is some loss of power, but this does not affect the performance greatly. In theory, a lossless matching network can be used, but since the power sensor operates over an extremely large frequency range (few KHz to 40 GHz is common), such matching is not useful.

The diode-based sensors do not show linearity comparable to the thermistor or thermocouple sensors. However, they can operate in the range 10^{-6} to 10^{-10} W, which the other types do not cover. Also, the diode sensors have a very small response time. Because of their reliance on thermal effects, the other types require some time (up to a few seconds) to settle to a valid reading, while the diode sensor does not have this limitation. As a result, it is the only sensor that can be used to display (for example) the variation of power in a modulated signal. So, the output from a diode sensor is frequently displayed on an oscilloscope, and a properly calibrated display (numeric or needle type) is not used.

3.2 Transmission Measurement

One basic measurement is of gain, which roughly relates to the ratio of output to input. Of course, when we try to quantify this, there are many definitions of *gain* that are encountered. We will start with $|S_{21}|$, which is a useful indicator of the performance of many circuits. So, the first and simplest microwave measurement that we will discuss is the measurement of $|S_{21}|$.

Figure 3.18 depicts the measurement of $|S_{21}|$ of a two-port component that is assumed to be linear and time invariant. The first point that should be clear is the impedance value with which the S-parameters were defined (see Section 2.1). Let us suppose that this is some R_0 (generalized S-parameters are not considered here).

If $R_s = R_L = R_0$, then $P_1 = |b_2|^2/(2R_L)$ and $P_0 = |a_1|^2/(2R_L) = |V_s|^2/(8R_L)$, which gives us

$$|S_{21}| = \sqrt{\frac{P_1}{P_0}} \tag{3.1}$$

So, the measurement of $|S_{21}|$ is simple enough and requires a source with impedance R_0 and a power meter with input impedance R_0. In practice, R_0 is almost always 50 Ω.

There is, of course, no guarantee that R_s and R_L will be both equal R_0, especially if a large frequency range is considered. In fact, these values will be in general complex. However, the value of $|S_{21}|$ evaluated using (3.1) is surprisingly insensitive to changes in R_s and R_L from the desired value of R_0. As an

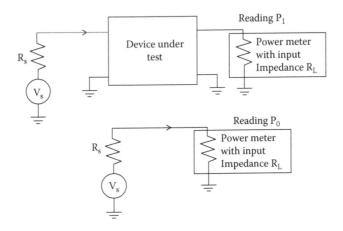

FIGURE 3.18
Basic measurement of $|S_{21}|$.

example, if the device under test (DUT) is a series capacitor of value 1 pF, then with $R_0 = 50\ \Omega$, the true value of $|S_{21}|$ is 0.53, while the value measured by using (3.1) if $R_s = 20\ \Omega$ and $R_L = 100\ \Omega$ is 0.6, and for $R_s = 40\ \Omega$ and $R_L = 60\ \Omega$ it is 0.532. So, the simple approach illustrated in Figure 3.18 is indeed very effective. To further enhance the accuracy, when R_s and R_L are significantly different from R_0, attenuators are used.

The circuit of a basic resistive attenuator is given in Figure 3.19.

The S-parameters of this component (reference R_0) are given by $S_{11} = S_{22} = 0$ and $S_{21} = S_{12} = A$, where A, the attenuation factor, can be set to any desired value of <1 by choosing R_1 and R_2 suitably. The circuit being resistive, the S-parameters are real and independent of frequency.

For example, a 10 dB attenuator has $A = 0.316$ (since $-20 \log 0.316 = 10$), and for a 20 dB attenuator, $A = 0.1$ (since $-20 \log 0.1 = 20$).

R_1 and R_2 can be evaluated from the equations

$$R_1 = R_0 \frac{1-A}{1+A}, \quad and \quad R_2 = R_0 \frac{2A}{1-A^2}$$

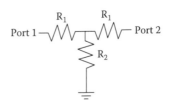

FIGURE 3.19
An attenuator using a resistive tee.

If a load is connected to port 2 of this attenuator, and we try to estimate the reflection coefficient looking into port 1, the incident wave will be attenuated by A before it reaches the load, and the reflected wave from the load will be again attenuated by A before it emerges from port 1 as the reflected wave. So, the reflection coefficient of the load is reduced by a factor A^2. Another way of putting this is that the impedance looking into port 1 is close to R_0, whatever the load. This is exactly the behavior we want. The price paid is, of course, the loss of power in the attenuator; if the DUT has a very low $|S_{21}|$, then adding attenuators may reduce the value of P_1 below what can be measured with the available power meter. This is the main limitation of this measurement.

Let us consider an example to see the effect of attenuators.

Example 3.1

A source with internal impedance of 30 ohms and a power meter with load impedance $50 + j20$ ohms are used to measure 50 ohm $|S_{21}|$ of a component that actually has the following S-parameters (mag and angle of S_{11}, S_{12}, S_{21}, and S_{22} with 50 ohm reference):

$$S_{11} = 0.85 \angle 129.2 \qquad S_{12} = 0.063 \angle -2.8 \qquad S_{21} = 0.628 \angle -40°$$
$$S_{22} = 0.71 \angle -168.1$$

Let us see the measured values, if (a) 6 dB three-resistor attenuators are used with both source and detector, and (b) no attenuators are used.

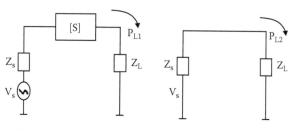

Figure 3.20 (a) with DUT   Figure 3.20 (b) without DUT

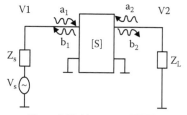

Figure 3.20 (c) waves at DUT

The measured $|S_{21}|_m$ is the ratio of the power delivered to the load with DUT (P_{L1}) to the power delivered to the load without DUT (P_{L2}), square-rooted.

$$|S_{21}|_m = \sqrt{\frac{P_{L1}}{P_{L2}}}$$

To derive P_{L1}, a_1, b_1, a_2, and b_2 are defined using $R_0 = 50\ \Omega$.
From the source (see Figure 3.20(c)) we have

$$\frac{V_s - V_1}{Z_s} = I_1$$

$$= \frac{V_s - (a_1 + b_1)}{Z_s} = \frac{a_1 - b_1}{R_o} \tag{3.2}$$

$$a_1 = \Gamma_s b_1 + \frac{R_o V_s}{R_0 + Z_s} \quad \left(\text{where } \Gamma_s = \frac{Z_s - R_0}{Z_s + R_o} \right) \tag{3.3}$$

$$a_1 = \Gamma_s b_1 + b_s \quad \left(\text{where } b_s = \frac{R_o V_s}{R_0 + Z_s} \right) \tag{3.4}$$

From the load:

$$\frac{V_2}{Z_L} = -I_2, \text{ i.e., } \frac{(a_2 + b_2)}{Z_L} = \frac{-(a_2 - b_2)}{R_o} \tag{3.5}$$

$$a_2 = \Gamma_L b_2 \quad \left(\text{where } \Gamma_L = \frac{Z_L - R_0}{Z_L + R_o} \right) \tag{3.6}$$

From S-parameters:

$$b_1 = S_{11}\, a_1 + S_{12}\, a_2 \tag{3.7}$$

$$b_2 = S_{21}\, a_1 + S_{22}\, a_2 \tag{3.8}$$

a_1, b_1, and a_2 can be obtained:

$$P_{L1} = |-I_2|^2 R_L = \left| \frac{b_2 - a_2}{R_o} \right|^2 R_L \tag{3.9}$$

$$P_{L1} = |b_2|^2 \left| \frac{1 - \Gamma_L}{R_o} \right|^2 R_L \tag{3.10}$$

Substitute Equations (3.4) and (3.6) into Equation (3.8):

$$b_2 = S_{21}(\Gamma_s b_1 + b_s) + S_{22}\Gamma_L b_2 \tag{3.11}$$

Substitute Equations (3.4) and (3.6) into Equation (3.7):

$$b_1 = S_{11}(\Gamma_s b_1 + b_s) + S_{12}\Gamma_L b_2 \tag{3.12}$$

Rearranging (3.12),

$$b_1 = \frac{S_{11}b_s + S_{12}\Gamma_L b_2}{1 - S_{11}\Gamma_s} \tag{3.13}$$

Substitute (3.13) into (3.11):

$$b_2 = \frac{S_{21}\,b_s}{1 - (S_{11}\Gamma_s + S_{22}\Gamma_L + S_{12}S_{21}\Gamma_L\Gamma_s) + S_{11}S_{22}\Gamma_s\Gamma_L} \tag{3.14}$$

where

$$b_S = \frac{R_o V_s}{R_0 + Z_s} \quad \text{(from Equation (3.4))}$$

Thus, by substituting Equation (3.14) in Equation (3.10), we have P_{L1}.
To calculate power delivered to load without DUT, see Figure 3.20(b).
We have

$$P_{L2} = \left|\frac{V_S}{Z_S + Z_L}\right|^2 R_L \tag{3.15}$$

Therefore, $|S_{21m}|$ can be obtained.

CASE I: WITH ATTENUATORS

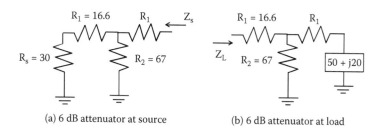

(a) 6 dB attenuator at source   (b) 6 dB attenuator at load

From Figure 3.21.

$$Z_s = 44.08\ \Omega$$

$$Z_L = 50.736 + j\,4.919$$

Calculate:

$$\Gamma_L = (Z_L - R_0)/(Z_L + R_0) = 0.0493 \angle 78.695°$$

$$\Gamma_S = (Z_S - R_0)/(Z_S + R_0) = 0.0628 \angle 180°$$

Using Equations (3.10), (3.14), and (3.15) calculate $|S_{21m}|$:

$$|S_{21m}| = 0.649$$

CASE II: WITHOUT ATTENUATORS

$$Z_S = 30 \ \Omega$$

$$Z_L = 50 + j20 \ \Omega$$

Calculate:

$$\Gamma_L = (Z_L - R_0)/(Z_L + R_0) = 0.196 \angle 78.69°$$

$$\Gamma_S = (Z_S - R_0)/(Z_S + R_0) = 0.25 \angle 180°$$

Using Equations (3.10), (3.14), and (3.15) calculate $|S_{21m}|$:

$$|S_{21m}| = 0.7115$$

which is of course not as close as case I, but may not be a bad estimate in many cases.

One assumption has been made above: it has been assumed in case I that the power displayed = some constant (say, C) times the power absorbed by Z_L (the attenuator + power meter). However, C is not known to be the same for the circuits in Figure 3.20(a, b). Students should work out the details to see if this assumption is correct.

A function similar to the attenuator, without loss of power, can be achieved if an isolator is available. This is a two-port component with S-parameters:

$$S_{11} = S_{22} = S_{12} = 0 \quad \text{and} \quad |S_{21}| \cong 1$$

If such isolators are connected after the source and before the power meter, then the measurement of $|S_{21}|$ will be accurate even if the source and power meter impedances are very different from R_0. Students should try to work out the mathmatical details of this circuit analysis. Unfortunately, isolators are somewhat expensive (and also bulky, since they usually use ferrites and permanent magnets), and are rarely used in this kind of measurement.

3.3 Reflection Measurement

Measurement of S_{11} is more difficult than measurement of S_{21}. We will start with a description of the slotted waveguide technique, which enabled measurement of the complex S_{11} (including phase). Before the advent of the vector network analyzer, this was a commonly used technique for such measurements.

Some preliminary information about the rectangular waveguide configuration is given in Figure 3.22.

The *inner dimensions* along the *x*- and *y*-directions are designated *a* and *b*, respectively. The wall thickness and outer dimensions are not of much concern. As is always the case in practice, only the TE_{10} mode is assumed to propagate. For this mode, the fields propagating in the *z*-direction are given by

$$\vec{E} = \hat{y}\, A\, sin\left(\frac{\pi x}{a}\right) e^{-j\beta z} \tag{3.16a}$$

and

$$\vec{H} = -\hat{x}\, A\left(\frac{\beta}{\omega\mu}\right) sin\left(\frac{\pi x}{a}\right) e^{-j\beta z} + \hat{z}\, jA\left(\frac{\pi}{a\omega\mu}\right) cos\left(\frac{\pi x}{a}\right) e^{-j\beta z} \tag{3.16b}$$

Here $\beta = \sqrt{\omega^2 \mu\varepsilon - \left(\frac{\pi}{a}\right)^2}$, $\omega = 2\pi f$, is the angular frequency, $\mu = \mu_0 = 4\pi \times 10^{-7}$ H/m, and $\varepsilon = \varepsilon_0 = 8.85 \times 10^{-12}$ F/m.

If we consider propagation along both *z*- and –*z*-directions, the electric field is given by

$$\vec{E} = \hat{y}\, A\, sin\left(\frac{\pi x}{a}\right) e^{-j\beta z} + \hat{y}\, B\, sin\left(\frac{\pi x}{a}\right) e^{j\beta z} \tag{3.17}$$

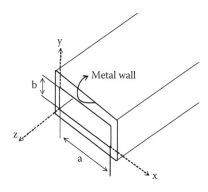

FIGURE 3.22
Co-ordinate system used for the rectangular metal waveguides.

The magnetic field can be written by using $\vec{H} = \frac{1}{-j\omega\mu} curl\ (\vec{E})$ but is not required here.

The ratio B/A is now defined as the reflection coefficient S_{11}. Notice that there is a deviation from the definition of S-parameters in terms of voltages and currents—we are not attempting to define *voltage* and *current* here. Instead, we are intuitively extending our understanding of S-parameters in terms of forward and backward waves. A more detailed discussion of the questions arising from this may be found in Itoh (1989).

Assume now that there is a technique (we will see the details later) for probing inside the waveguide and obtaining a voltage signal that is proportional to the peak (i.e., maximized over x) electric field inside the waveguide at any specified z-value. This is done without substantially disturbing the fields described by (3.17), so the voltage is expected to be small.

This voltage is a function of z, given by

$$V(z) = K(A\ e^{-j\beta z} + B\ e^{j\beta z}) \tag{3.18}$$

The constant K, as mentioned above, is small. Of course, any such voltage has an associated impedance, which in this case is assumed to be independent of z. So, the Thevenin equivalent of this probed output is given in Figure 3.23.

The strength of the probed voltage is measured with a power meter (usually the diode sensors are used). Often an impedance matching network (or tuner) is inserted between the probe output and the power meter to match Z_s to the power meter input impedance.

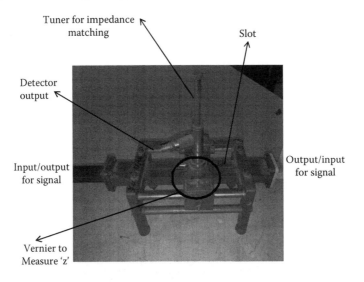

FIGURE 3.23
(a) 6 dB attenuator at source. 6 dB attenuator at detector.

We will try to relate the power meter reading to A and B. Assume that peak values, and not root mean square (rms), are used.

The power available (P_{av}) from $V(z)$ is $|V(z)|^2/(8\,Re(Z_s))$, and the displayed power is proportional to this. So, we have to evaluate the magnitude of the expression in (3.18).

Let $S_{11} = re^{j\theta}$, $B = S_{11}$ and $A = A\,re^{j\theta}$. Then

$$P_{av} = \frac{1}{8Re(Z_s)}|KA|^2\,|e^{-j\beta z} + re^{j\theta}e^{j\beta z}|^2$$

$$= \frac{1}{8Re(Z_s)}|KA|^2\,|\cos(\beta z) + r\cos(\beta z + \theta) + j\,(-\sin(\beta z) + r\sin(\beta z + \theta))|^2$$

$$= \frac{1}{8Re(Z_s)}|KA|^2\,[1 + r^2 + 2r\cos(2\beta z + \theta)]$$

So,

$$P_{av}(z) \propto [1 + r^2 + 2r\cos(2\beta z + \theta)] \tag{3.19}$$

Referring to Figure 3.22, and assuming that the unknown load that gives rise to reflection is connected at $z = 0$, we see that z-values of interest are all negative. Now suppose that the minimum power displayed is observed at $z = -d_{min}$. From (3.19), minimum power will be observed when

$$\cos(2\beta z + \theta) = -1$$

or

$$-2\beta d_{min} + \theta = (2n + 1)\pi$$

or

$$\theta = (2n + 1)\pi + 2\beta d_{min}$$

The numerically smallest value of θ is given by $\theta = -\pi + 2\beta d_{min}$. The other values differ by multiples of 2π, so the choice is a minor matter.

Finally, suppose we denote the ratio of maximum to minimum powers displayed (as we change z) by R. Being a ratio, this is also the ratio of maximum and minimum powers available. $\sqrt{R}$ is given the name *voltage standing wave ratio* (VSWR) and is denoted by S. Like most ratio-based measurements, we can expect that S is known to a high degree of accuracy.

From (3.19),

$$S = \frac{1+r}{1-r}$$

that is,

$$r = \frac{S-1}{S+1}$$

So, both magnitude and phase of reflection coefficient (r and θ) have been determined. All this is of course subject to the possibility of such probing. Let us see how this is done.

3.3.1 The Slotted Waveguide

The waveguide is a fully shielded guiding structure, and in order to carry out any probing, one cannot avoid disturbing the fields given by (3.16) at least to a small extent. The above description implicitly assumes that:

1. A cut or an opening of some sort is made in the guide wall. Since measurements over a continuous range of z-values are planned, this opening will take the form of a long slot.
2. We want the probed signal to depend on z alone and not x or y. So, the slot has to be parallel to the z-axis.

The only questions that remain to be answered are: Will the slot be on a broad wall or a narrow wall? Where (at what offset from the wall center) will the slot be located?

To answer these, we notice that the effect of cutting a narrow slot in a metal is that the currents flowing on the metal surface that are perpendicular to the slot are stopped. So, a narrow slot that is entirely parallel to the current direction in the undisturbed waveguide will have no effect. The surface current density on a perfect conductor is usually denoted by J_s (which has units A/m; to distinguish it from the usual current density J with units A/m², see any textbook on electromagnetics [e.g., Pozar, 1989] and is connected to the fields by

$$\vec{J}_s = n \times \vec{H} \tag{3.20}$$

Here, n is the outward normal, i.e., directed from the metal (where there is no field) to the air (where fields are present).

Using this relation at the top wall ($y = b$ and $n = -y$), we get

$$\vec{J}_{s,\,top} = -\hat{z}\, A \left(\frac{\beta}{\omega\mu} \right) \sin\left(\frac{\pi x}{a} \right) e^{-j\beta z} - \hat{x}\, jA \left(\frac{\pi}{a\omega\mu} \right) \cos\left(\frac{\pi x}{a} \right) e^{-j\beta z} \tag{3.21}$$

At the center of the top wall ($x = a/2$), we see that the current density is entirely z-directed, and we can indeed cut a narrow z-directed slot without disturbing it.

On the narrow wall at $x = 0$, $n = x$, and we get

$$\vec{J}_{s,\,side} = -\hat{y}\, jA \left(\frac{\pi}{a\omega\mu} \right) e^{-j\beta z} \tag{3.22}$$

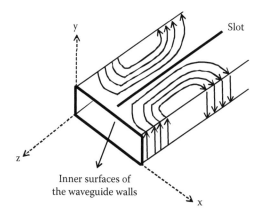

FIGURE 3.24
Current paths on the inner walls of a waveguide for the TE10 mode.

This is entirely *y*-directed and constant over the wall. So, there is no way to cut a z-directed slot that does not intersect the current path. This will obviously hold true for the other narrow wall as well.

Evidently, the slot has to be cut in the center of the top wall. Of course, the bottom wall will do as well (it is just a matter of rotating the waveguide 180° about the z-axis—top and bottom are fully interchangeable).

The actual current paths in the TE_{10} mode are sketched in Figure 3.24. As usual in such plots, this is to be interpreted as a snapshot at a particular time instant—so the expressions in (3.21) and (3.2) should be converted to time domain and any convenient *t* (say, *t* = 0) should be selected. The curves in Figure 3.23 are drawn so that the direction of J_s (real-valued in time domain now) is tangential to them at any point on the walls. It is obvious by inspection that the center of the broad wall is the only choice for cutting a slot.

An actual slotted waveguide assembly is shown in Figure 3.25. What is missing in Figure 3.25 is the actual probe, i.e., what exactly goes into the slot. The simplest type of probe is an extended inner conductor of a thin co-axial cable. This is illustrated in Figure 3.26.

When the probe is positioned at some location $z = z_0$, it is *assumed* that the output $V(z = z_0)$, referring to Figure 3.23, is dependent only on the electric field at z_0. This is only an approximation in practice. Even if the probe tip (the inner conductor in Figure 3.26) were very thin, the field lines in some neighborhood of the probe would be distorted, and it is not possible to select the field only at z_0. This is a limitation that is fundamental to the approach—any correction is not really practical. Still, this technique was good enough to be widely used before network analyzers became common. We have not discussed the impedance matching section following the probe. Information on this may be found from Maury Corp. (2013).

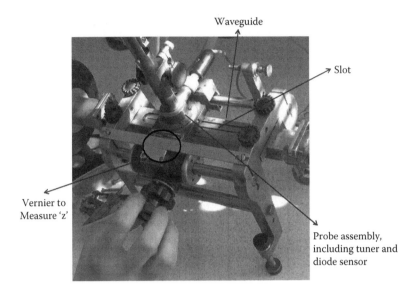

FIGURE 3.25
A slotted waveguide bench.

Also, we have assumed that the microwave source is a continuous wave (CW) in nature. This need not be so, and often the source is ON/OFF modulated at ~1 KHz. This permits us to use a high gain ac-coupled amplifier after the power sensor to boost the sensitivity, instead of relying on chopper circuits as is incorporated in power meters that must cater to CW measurements. Since the absolute value of power is not a concern in this type of measurement, a high-performance properly calibrated power meter is not required.

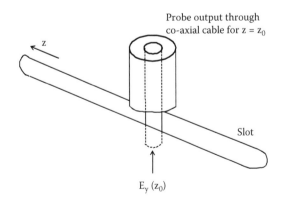

FIGURE 3.26
Enlarged view of a typical probe configuration.

In the measurement of S_{11} above, it was mentioned that the location of the minimum probe output was noted. We can in principle consider the location of the maximum as well. However, these is a difference, most noticeably when the VSWR is high. As an extreme case, take $S_{11} = 1$. Now (3.19) becomes

$$P_{av}(z) \propto [1 + \cos(2\beta z)]$$

Let us tabulate the values of the expression $[1 + \cos(\theta)]$ for several values of θ.

θ (radians)	$1 + \cos(\theta)$	$10 \log(1 + \cos\theta)$
3.14	1.268×10^{-6}	−58.97
3.1	8.648×10^{-4}	−30.6
0	2	3.01
0.04	1.9992	3.008

If we arbitrarily assign units of mW to the second column (so the third column will be in dBm), it is obviously easy to distinguish between −30.6 and −58.97 dBm with a power meter. So, while homing in to a minimum, we are able to resolve the phase to 0.04 radians (which translates to roughly a 0.1-mm linear resolution at the X band, which is typical for a vernier). On the other hand, near the maximum we will be forced to distinguish between 3.01 and 3.008 dBm, which is not practical, even with very sensitive equipment. This is the reason why slotted waveguide measurements focus on minimum values (and try to work with large reflections) and not maximum.

Also, we have seen that power sensors are more accurate (linear relation between RF power and output dc voltage) at lower powers—this was also an argument for focusing on minimum rather than on maximum in the pre-computer days, when a large set of calibration data could not be readily used with a power sensor.

Special techniques have evolved (Sucher and Fox, 1963) to further augment the capability of the slotted waveguide, such as measuring high VSWR values accurately or measuring low VSWR values of lossless components (such as filters in the pass-band). There is also a co-axial version of the slotted waveguide (http://www.hpmemory.org/wa_pages/wall_a_page_06.htm) that works on very similar principles.

Today, of course, these setups have been replaced by vector network analyzers. They now mostly find application in educational and training purposes.

3.3.2 Use of the Directional Coupler

Measurement of the reflection coefficient can be also accomplished using a directional coupler. The setup is shown in Figure 3.27. The goal of this is measurement of $|S_{11}|$—it is a scalar measurement. To avoid confusion, let us

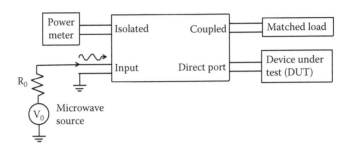

FIGURE 3.27
Reflection measurement using a directional coupler.

use the symbol Γ for the reflection coefficient of the DUT. S_{ij} will be used for the directional coupler.

Let us number the ports: 1, input; 2, direct port; 3, coupled port; and 4, isolated port. Obviously the designations *coupled*, *isolated*, and *direct* are all with reference to the input, so if the input is connected to a different port, the other designations will also change.

If the coupler is assumed to be ideal, then it has S-parameters (in some reference R_0, typically 50 Ω):

$$\begin{bmatrix} 0 & T & C & 0 \\ T & 0 & 0 & C \\ C & 0 & 0 & T \\ 0 & C & T & 0 \end{bmatrix}$$

where T is the direct-port coupling and $|T| \sim 0.9$, and C is the coupling and $|C| \sim 0.1$ (but is fairly well known). Further, there is no reflection from the coupled port, and we will also assume that the source and power meter impedances are R_0 (or there is an isolator or attenuator with these components, as discussed earlier).

So, we get the relations $a_3 = 0$ and $a_4 = 0$. Using the fourth row of the S-matrix of the coupler,

$$b_4 = Ca_2 = C\,(\Gamma b_2)$$

Using the second row of the coupler S-matrix,

$$b_2 = Ta_1$$

$$\Rightarrow b_4 = \Gamma CT\, a_1 \tag{3.23}$$

Now

$$a_1 = (V_1 + R_0\, I_1)/2 = V_0/2 \tag{3.24}$$

Finally,

$$\text{Power displayed} = [|-I_4|^2 R_0/2] = |b_4|^2/(2R_0) \qquad (3.25)$$

Combining (3.23), (3.24), and (3.25), power displayed (say, P) = $[V_0^2/(8R_0)]$ $|CT|^2 |\Gamma|^2$.

If we replace the DUT with a short that has reflection coefficient -1, the power displayed (call this P_0) will be $[V_0^2/(8R_0)] |CT|^2$.

So, we get, finally

$$|\Gamma| = \sqrt{\frac{P}{P_0}}$$

This is a simple and effective procedure for obtaining the reflection coefficient magnitude, *but* it depends on the availability of a good coupler with S-parameters close to the ideal ones and matched source and power meter. As we have seen earlier, this is doubtful. Let us see the effect of nonideality on this measurement.

Example 3.2

A 20 dB symmetrical coupler is used to measure reflection coefficient with a short. The only nonideality is that the coupler directivity is 16 dB. If the true $|\Gamma| = 0.5$, what is the range of the measured values?

Power meter, source, etc., are matched to 50 Ω.

Referring to Fig. 3.28, Coupler S-parameters:

$$[S] = \begin{bmatrix} 0 & T & C & I \\ T & 0 & I & C \\ C & I & 0 & T \\ I & C & T & 0 \end{bmatrix}$$

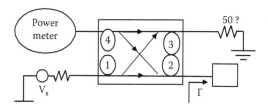

FIGURE 3.28
Circuit for calculation of reflection coefficent.

where $|C| = 0.1$ for 20 dB.

$$\text{Isolation} = \text{Directivity} + \text{Coupling} = 36 \text{ dB}$$

$$|I| = 1/\sqrt{4000}$$

Assume lossless system and

$$|C|^2 + |T|^2 + |I|^2 = 1 \tag{3.26}$$

$$|T| = 0.99486$$

The true $|\Gamma| = 0.5$.

$$a_2 = \Gamma\, b_2 \tag{3.27}$$

$$a_3 = 0 \tag{3.28}$$

$$a_4 = 0 \tag{3.29}$$

Also, we have coupler *S*-parameters:

$$
\begin{bmatrix} b_1 \\ b_2 \\ b_3 \\ b_4 \end{bmatrix}
=
\begin{bmatrix} 0 & T & C & I \\ T & 0 & I & C \\ C & I & 0 & T \\ I & C & T & 0 \end{bmatrix}
\begin{bmatrix} a_1 \\ a_2 \\ a_3 \\ a_4 \end{bmatrix}
$$

$$b_1 = Ta_2 \tag{3.30}$$

$$b_2 = Ta_1 \tag{3.31}$$

$$b_3 = Ca_1 + Ia_2 \tag{3.32}$$

$$b_4 = Ia_1 + C\,a_2 \tag{3.33}$$

$$b_4 = (I + C\Gamma T)a_1 \tag{3.34}$$

With device under test:

Power meter reading, $P_1 = |b_4|^2/(2R_0)$

$$P_1 = |I + CT\Gamma|^2 |a_1|^{2}/(2R_0) \tag{3.35}$$

With short, $\Gamma = -1$:

$$P_2 = |I - CT|^2 |a_1|^{2}/(2R_0) \tag{3.36}$$

Let $|\Gamma_m| = \sqrt{\frac{P_1}{P_2}}$, where Γ_m is the measured value Γ.

$$|\Gamma_m| = \left|\frac{I + CT\Gamma}{I - CT}\right| = \left|\frac{1 + CT\Gamma/I}{1 - CT/I}\right|$$

The magnitudes of I, C, Γ, and T are known.
For $|\Gamma_m|_{max}$, $\angle \frac{CT\Gamma}{I} = 0$ and $\angle \frac{CT}{I} = 0$,

$$|\Gamma_m|_{max} = 0.783$$

For $|\Gamma_m|_{min}$, $\angle \frac{CT\Gamma}{I} = \pi$ and $\angle \frac{CT}{I} = \pi$,

$$|\Gamma_m|_{min} = 0.2942$$

The directional coupler-based reflection measurement forms a link to the modern vector network analyzer, which will be discussed in Chapter 4.

3.4 Conclusion

In this chapter we described some traditional techniques of microwave measurements. There has been a huge improvement in this field, particularly after the advent of high-speed analog-to-digital convertors (ADCs) and associated processing capability. Some of the concepts described here, such as power sensing with diode, however, still continue to be valid. The directional coupler-based reflection coefficient measurement forms a link to the modern network analyzer.

Problems

1. What impedances can be seen looking into port 1 of a 10 dB resistive attenuator for different loads at port 2?

2. An ideal power meter is designed with an input impedance of 30 Ω. What will it read if excited by a 50 Ω source with available power $= 1$ W? How will the reading change if an ideal isolator (designed for 50 Ω) precedes the power meter?

3. A 20 dB symmetrical coupler is used to measure reflection coefficient with a short. The only nonreality is that the coupler directivity is 16 dB. If the measured $|S_{11}| = 0.5$, what is the maximum and minimum possible values of the true $|S_{11}|$?

4. An X band (WR90) waveguide is excited at 10 GHz and shorted at the other end. When positioned at any z_0, the probe picks up a signal proportional to the average electric field in a z-range $z_0 - 1$ mm to $z_0 + 1$ mm. What will be the pattern of the displayed power as a function of z?

5. Referring to Equation (3.17), suppose we define Ab and Bb (b is the waveguide height) as amplitudes of voltage waves on a transmission line, and $(\omega\mu/\beta)$ as characteristic impedance Z_c. How can we define current, and how do we interpret the input impedance $Z_c \frac{1+S_{11}}{1-S_{11}}$?

References

Agilent. *Fundamentals of RF and Microwave Power Measurements. Part 2. Power Sensors and Instrumentation.* Agilent Technologies Application Note 1449-2, 2006.

Itoh, T. *Numerical Techniques for Microwave and Millimeter-Wave Passive Structures,* John Wiley & Sons, New York, 1989.

Malik, R.J., T.R. Aucoin, and R.L. Ross. Planar-Doped Barriers in GaAs Molecular Beam Epitaxy. *Electronics Letters,* 1G(22), 1980.

Maury Corp. Wide Matching Range Slide Screw Tuners. Catalog, November 2013.

Millman, J., and A. Grabel. Microelectronics. McGraw-Hill, New York, 1987.

Pozar, D.M. *Microwave Engineering,* John Wiley & Sons, New York, 1989.

Sucher, M., and J. Fox, Eds. *Handbook of Microwave Measurements,* Vol. 1, 3rd ed. Polytechnic Press, Brooklyn, NY, 1963.

Szente, P.A., S. Adam, and R.B. Riley. Low-Barrier Schottky-Diode Detectors, *Microw. J.,* 1976, 19(2).

4

Vector Network Analyzer

The vector network analyzer (VNA) is the most important instrument in microwave measurements today. Its primary function is the measurement of S-parameters (remember that these are defined for linear time-invariant systems). Starting from the rudimentary setup using a directional coupler and some power sensors described in Chapter 3, this instrument has evolved into the workhorse of any microwave laboratory, and is today moving into measurement of nonlinear properties of components, and the frequency of operation today entering the THz region.

Here we will discuss the principle of operation of this instrument in its simplest form.

4.1 Enhancement of Scalar Measurement

In Chapter 3, we saw how a directional coupler can be used to measure $|\Gamma|$ (obviously, this is the same as measuring S_{11} of a component, but we will use the symbol Γ here to avoid confusion with S_{11} of the directional coupler). Let us take a relook at that procedure.

The measurement setup is shown in Figure 4.1.

The idea is that the input to the power meter comes purely from reflection from the device under test (DUT), and hence the reading is proportional to $|\Gamma|^2$. The constant of proportionality can be eliminated by taking a second reading with a known reflection coefficient replacing the DUT, and taking the ratio of the two readings. The known reflection coefficient is most often a short, since it is usually practical to fabricate a short circuit that retains the property of $\Gamma = -1$ at the highest frequencies used in measurements. The basis of this procedure is the coupler behavior, which can be conveniently captured through its S-parameters as discussed in Chapter 3:

$$[S] = \begin{bmatrix} 0 & T & C & 0 \\ T & 0 & 0 & C \\ C & 0 & 0 & T \\ 0 & C & T & 0 \end{bmatrix} \tag{4.1}$$

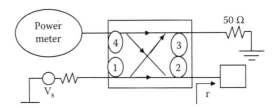

FIGURE 4.1
Procedure for measurement of $|\Gamma|$.

Complications arise when deviations from ideality in this setup are considered. These are:

1. The source and power meter impedances may not be 50 Ω (the reference in defining Γ).
2. The coupler S-parameters may deviate from (4.1).

The termination at port 3 can also be different from 50 Ω, but this deviation is usually insignificant.

Let us consider the effect of one particular nonideality: suppose that the coupler isolation is not 0, but some I (of course, C, T, and I are complex).

Now the coupler S-matrix becomes

$$[S] = \begin{bmatrix} 0 & T & C & I \\ T & 0 & I & C \\ C & I & 0 & T \\ I & C & T & 0 \end{bmatrix} \tag{4.2}$$

It was shown in Chapter 3 that

$$\sqrt{\frac{P_1}{P_2}} = \left| \frac{I + CT\Gamma}{I - CT} \right| \tag{4.3}$$

where P_1 is the power reading with the DUT connected and P_2 is the reading where the DUT is replaced with a short.

A small modification gives

$$\sqrt{\frac{P_1}{P_2}} = \left| \frac{(I/CT) + \Gamma}{\left(\dfrac{I}{CT}\right) - 1} \right|, \quad \text{or} \quad \left| \Gamma + \left(\frac{I}{CT} \right) \right| = \sqrt{\frac{P_1}{P_2}} \left| \left(\frac{I}{CT} \right) - 1 \right| \tag{4.4}$$

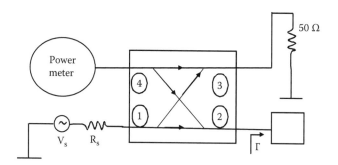

FIGURE 4.2
Directional coupler measurement with unmatched source.

Now, if a 50 Ω transmission line of some known electrical length θ is inserted between port 2 and the DUT (not connected when DUT is replaced with the short), then (4.4) will change to

$$\left| \Gamma e^{-2j\theta} + \left(\frac{I}{CT} \right) \right| = \sqrt{\frac{P_1}{P_2}} \left| \left(\frac{I}{CT} \right) - 1 \right|, \text{ or}$$

$$\left| \Gamma + \left(\frac{e^{-2j\theta}}{CT} \right) \right| = \sqrt{\frac{P_1}{P_2}} \left| \left(\frac{I}{CT} \right) - 1 \right| \tag{4.5}$$

Since Γ is the only unknown in (4.5), this is the equation of a circle when Γ is plotted in the complex plane. Moreover, the center and radius of this circle will change when different values of θ are used (remember that P_1 will change in this case). The intersection of three or more such circles will give the *complex value of* Γ. So, a poor directional coupler actually gives us more information than we wanted when we devised this procedure.

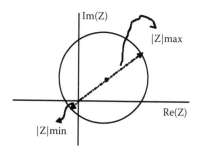

FIGURE 4.3
Illustration of maximum and minimum |Z| when Z lies on a circle.

Actually, what is happening is that the transmission line changes the phase of the reflected signal, and hence there is constructive or destructive interference of this reflected signal with the leaked signal from port 1 to 4—leading to changes in the power meter reading. In practice, of course, it is not likely that limited isolation is the only problem and everything else is ideal—so a more detailed study will be required. Further investigation of this technique leads to the concept of interferometry (Engen and Hoer, 1992).

Before moving on, let us see an example relating to source impedance variation.

Example 4.1

A 10 dB symmetrical ideal (in all respects) directional coupler is designed for use in a 50 Ω system, used with a 100 Ω source, ideal short, ideal load (50 Ω), and ideal reflection-less power meter (i.e., 50 Ω input impedance), to measure the reflection coefficient (magnitude) of a component normalized to 50 Ω. If the measured value is 0.5, what can you conclude about true value?

Coupler S-parameters:

$$[S] = \begin{bmatrix} 0 & T & C & 0 \\ T & 0 & 0 & C \\ C & 0 & 0 & T \\ 0 & C & T & 0 \end{bmatrix} \quad \text{where } |C|^2 = 0.1 \text{ for 10 dB}$$

and $|C|^2 + |T|^2 = 1$.

Again, write the relation between a_i and b_i (using reference $R_0 = 50\ \Omega$). (Note: Be careful about the meaning of forward and backward wave; it depends on whether we are referring to the coupler or to Γ.)

$$\frac{a_1 - b_1}{R_0} = I_1 = \frac{V_S - V_1}{Z_S} = \frac{V_S - (a_1 + b_1)}{Z_S} \tag{4.6}$$

$$\Rightarrow a_1 = \Gamma_S b_1 + \frac{R_0 V_S}{R_0 + Z_S}, \text{ where } \Gamma_S = \frac{Z_S - R_0}{Z_S + R_0} \tag{4.7}$$

Also

$$a_2 = \Gamma b_2, a_3 = 0, a_4 = 0 \tag{4.8}$$

We have coupler S-parameters:

$$\begin{bmatrix} b_1 \\ b_2 \\ b_3 \\ b_4 \end{bmatrix} = [S] \begin{bmatrix} a_1 \\ a_2 \\ a_3 \\ a_4 \end{bmatrix} \tag{4.9}$$

Using coupler S-parameters, we have

$$b_1 = Ta_2, \ b_2 = Ta_1, \ b_3 = Ca_1, \ b_4 = Ca_2 \tag{4.10}$$

Using Equations (4.10) and (4.8),

$$b_4 = C\Gamma b_2 \quad \text{and} \quad b_1 = T\Gamma b_2$$

Substitute $a_1 = b_2/T$ and $b_1 = T\Gamma b_2$ in (4.7), and solving for b_2 gives

$$b_2 = \frac{\left(\dfrac{V_S R_0}{Z_S + R_0}\right)}{\dfrac{1}{T} - T\Gamma\Gamma_S}, \quad \text{and} \quad b_4 = \frac{\left(\dfrac{V_S R_0}{Z_S + R_0}\right)C\Gamma}{\dfrac{1}{T} - T\Gamma\Gamma_S} \tag{4.11}$$

Using the relations, find the power meter reading:

$$P_1 = |b_4|^2 R_0 = \left| \frac{\left(\dfrac{V_S R_0}{Z_S + R_0}\right)C\Gamma}{\dfrac{1}{T} - T\Gamma\Gamma_S} \right|^2 R_0 \tag{4.12}$$

For a reference, replace the component by a short, power meter reading = P_2.

To calculate P_2, replace $\Gamma = -1$ in (4.12), and then take the ratio to get the measured Γm:

$$|\Gamma_m| = \sqrt{\frac{P_1}{P_2}} = \left| \frac{-\Gamma(1 + T^2\Gamma_S)}{1 - T^2\Gamma\Gamma_S} \right| \tag{4.13}$$

This results in an equation like:

$$|\Gamma_m| = \left[\frac{P + Q\Gamma}{P' + Q'\Gamma} \right],$$

where P, Q, P', and Q' are all known if T and C are known. Assume $\angle T = -\pi/2$ and $\angle C = -\pi$. From coupler relations, we have $T = 0.9486 \angle -\pi/2$, and we know that $|\Gamma_m| = 0.5$ and $\Gamma_S = 1/3$.

Substitute $\Gamma = Z = x + jy$, in (4.13), and using these numbers, we have after some calculations:

$$x^2 + y^2 - 0.32x - 0.5347 = 0$$

The above equation is a circle where

$$\text{Center} = (0.16, 0) \quad \text{Radius} = 0.748$$

So,

$$|Z|\max = |center| + |radius| = 0.908$$

$$|Z|\min = |\ |center| - |radius|\ | = 0.588$$

Therefore, $0.908 < |\Gamma| < 0.588$

4.2 Basic Vector Measurements

The real breakthrough in network analysis came with the introduction of phase measurement. While this can theoretically be done in the analog domain, it really becomes a powerful technique when the signals under consideration are digitized and processed digitally. Since this cannot be done at microwave frequencies, some form of frequency downconversion will be required. Let us see how this affects the phase.

Assume that we want to measure the phase difference between two signals: $A \cos(\omega_R t + \theta)$ and $B \cos(\omega_R t + \phi)$, where ωR is the frequency that is the same for these signals, and is likely to be in the GHz region. The obvious implication is that these signals came from the same source.

Now suppose that these signals are downconverted by mixing with a common LO as shown in Figure 4.4.

Assuming that ω_L is just below ω_R, we get the outputs:

Output 1 = low-frequency component of $A \cos(\omega_R t + \theta) \cos(\omega_L t + \alpha)$

$$= KA \cos((\omega_R - \omega_L)t + \theta - \alpha)$$

Output 2 = low-frequency component of $B \cos(\omega_R t + \theta) \cos(\Omega_L t + \alpha)$

$$= KB \cos((\omega_R - \omega_L)t + \phi - \alpha)$$

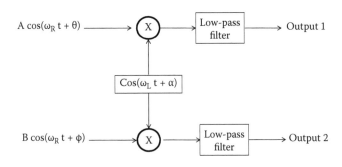

FIGURE 4.4
Downconversion for phase measurement.

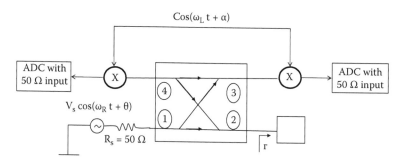

FIGURE 4.5
Measuring complex reflection coefficient.

The phase difference between these two is $(\theta - \phi)$, which is exactly the same as for the original signals, although the frequency of the signals is now very low (may be in KHz). This can be digitized and the phase difference $(\theta - \phi)$ determined by digital processing. This remains true even if α has a small time variation, i.e., the LO is not synchronized with the input frequency. If, however, there is an asymmetry in the LO connection to the two mixers, or the mixers are not identical, then, of course, there will be a problem—now there will be different α_1 and α_2 in the output signals instead of the same α. Even this can be remedied by a onetime measurement with a known difference in the input phases—the obvious case being identical inputs that can be generated with a symmetrical power splitter. Now the difference $(\alpha_1 - \alpha_2)$ will be known and the circuit is ready for measurements with new $(\theta - \phi)$; the results will be corrected by adding the known $(\alpha_1 - \alpha_2)$. A similar correction may be made for the gain factor K in the expressions above (which may become K_1 and K_2).

With this background, we can now see how the setup in Figure 4.1 can be modified for measurement of complex Γ. This is shown in Figure 4.5. It is assumed that each mixer has a low-pass (or band-pass centered at $\Omega_R - \Omega_L$) filter connected to its output—this is not separately shown.

Assuming no reflection from the mixers or analog-to-digital convertors (ADCs), we can conclude that the ADC at port 3 records a waveform with amplitude $= K\,|a_1\,C|$, while the ADC at port 4 records a waveform with amplitude $= K\,|a_1\,C\,T\,\Gamma|$. So, the amplitude ratio gives us $|T\Gamma|$. Also, the phase difference between these waveforms is $\angle T\Gamma$. So, if the complex T is known, Γ can be determined. Additionally, if Γ is known (e.g., a short gives $\Gamma = -1$), then the complex T can be determined, and now the system is calibrated to measure an unknown Γ. A further extension to enable transmission measurement for a two-port DUT is shown in Figure 4.6.

Here Γ and G for the DUT are actually identical to its S_{11} and S_{21}, assuming its port 1 is connected to the coupler (different symbols are used only to avoid confusion with the coupler S-parameters). Now the ADC3 will

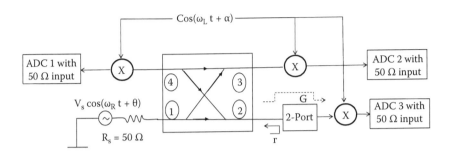

FIGURE 4.6
Measurement of transmission and reflection.

record a waveform with magnitude $= K\ |a_1\ T\ G|$. Comparing this to ADC2, which shows a magnitude $K\ |a_1\ C|$, we get $|GT/C|$. Also, the phase difference between the waveforms recorded by ADC3 and ADC2 is $\angle G + \angle T - \angle C$. Effectively, we have measured the complex GT/C. Assuming that T/C is known, G can be determined. Alternatively, if G is known (for example, a direct connection of port 2 of the coupler with the mixer; thus, removing the DUT gives $G = 1$), then T/C can be determined. This setup is still found today as the transmission/reflection test set in some network analyzers. An implicit assumption in the above is that the ADCs are all synchronized with the same clock to enable phase comparison.

4.3 Architecture of the Vector Network Analyzer

The architecture of a two-port vector network analyzer is shown in Figure 4.7—the key components here are the directional couplers. Regarding the symbol used for them, if any of the four ports is considered as the input, then the through port is indicated by a horizontal arrow, while the coupled port is indicated by a tilted arrow.

Conceptually, this is very similar to the setup shown in Figure 4.6. The main difference is the addition of a second coupler and an SPDT (single-pole double-throw) switch to enable measurement of all four S-parameters without changing any connection manually. The setup of Figure 4.6 can also do the job, but there we will have to manually reverse the ports of the DUT.

When the switch is in the position shown, a microwave signal at frequency f_R is supplied to directional coupler 1 (signal a_1). This is routed to port 1 of the DUT using the through- or direct-port path of the coupler; hence, almost all the signal reaches the DUT. The DUT will generate b_1 and b_2. As can be seen, these are downconverted and digitized by ADC2 and ADC3, respectively. ADC1 is separately configured to digitize the excitation, which can be a_1 or

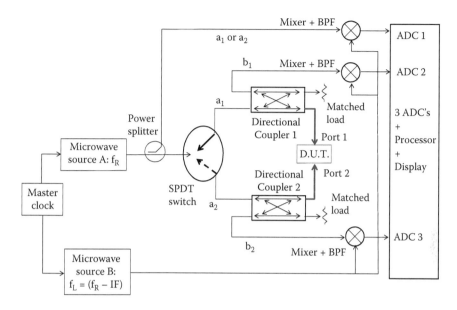

FIGURE 4.7
Architecture of a vector network analyzer.

a_2, depending on the switch position (the solid arrow for a_1 and the dashed arrow for a_2). This switch is of the absorptive type—so in the present state directional coupler 2 will see a 50 Ω load looking into the switch. There are also variable gain IF amplifiers preceding the ADCs (not shown here) to cater to situations where b_1 or b_2 is very small. The ADCs naturally have to be synchronized by a single clock that can be the 10 MHz master clock used by the microwave sources.

In the present configuration, the ADCs can operate as described earlier and obtain

$$S_{11} = b_1/a_1 \quad \text{and} \quad S_{21} = b_2/a_1$$

The switch can next be reversed so that the excitation is a_2, applied to directional coupler 2. Now the other two parameters are determined:

$$S_{12} = b_1/a_2 \quad \text{and} \quad S_{22} = b_2/a_2$$

The main excitation frequency f_R (which can sweep over a range as high as 0.1 to 40 GHz with resolution < 1 Hz) is generated by microwave source A, as shown, and source B, which supplies the LO signal to the mixers, is always kept a fixed IF below f_R (actually the downconversion typically takes place in multiple stages, as will be discussed in connection with spectrum analyzers in Chapter 5, but the overall effect is the same). Both of these are actually generated from a fixed oscillator/clock of very high stability (typically a crystal oscillator). ten MHz has emerged as the standard value of this reference. Generation of signal in this way will be discussed in Chapter 7.

Since the mixer output is always at a fixed value, the IF gain amplifiers can be designed to function with very low noise and well-controlled gain at precisely that frequency. Such a single-frequency operation results in very low noise (noise is known to be proportional to bandwidth), and consequently the network analyzer can accurately display S-parameters that are ≤ 100 dB in magnitude.

Directional couplers that maintain a good directivity over a band as huge as 0.1 to 40 GHz (and beyond) have been perfected by the few companies manufacturing such equipment. This is the capability that identifies a successful VNA manufacturer—the rest of the components are well understood. Exact details about these couplers are not available publicly. One approach that has probably been used is centered around two air-dielectric co-axial lines that are parallel and closely spaced, with a thin metallic wall separating them. A suitably shaped aperture in this wall provides broadband coupling.

4.4 Network Analyzer Calibration

It is clear from the discussion so far that unless proper calibration is done, the imperfections in the coupler, the reflection from the SPDT switch, etc., will prevent any accurate measurement of the DUT S-parameters. At most, such an uncalibrated measurement may tell us if the DUT is working or damaged. Many ways of calibrating vector network analyzers have been developed (LRL, LRM, SOLR, etc.—check Dunsmore (2012) for details); here we will only discuss the simplest two. Both depend on a mathematical model of the network analyzer.

4.4.1 SOLT Calibration (One-Port)

SOLT stands for short-open-load-through; another name for this type of calibration is TOSM (through-open-short-match). These terms refer to calibration standards—components whose S-parameters are accurately known in advance, and looking at their measured S-parameters, errors in the network analyzer may be quantified and corrected for. The short with $S_{11} = -1$ was already used as such a standard in Section 4.2.

Let us start with the mathematical model for the setup in Figure 4.4. This approach uses signal-flow graphs and has been popularized by Agilent Technologies (Agilent 1287-3). The following quantities are now defined:

a_1: The incident signal at the DUT
b_1: The reflected signal from the DUT

a_0: The output of the mixer connected to port 3 of the coupler

b_0: The output of the mixer connected to port 4 of the coupler

Being low-frequency signals, a_0 and b_0 are also considered to be identical to what is digitized and processed, to emerge as two complex numbers.

If everything were ideal, we would have $a_0 = K_a a_1$ and $b_0 = K_b b_1$. Now our objective of measuring (b_1/a_1) would have been satisfied if we knew the complex number (K_b/K_a). This could be achieved by measuring the reflection coefficient of a short, which is known to be –1. For this we would have $b_0/a_0 = (K_b/K_a)(b_1/a_1) = -(K_b/K_a)$, and (b_0/a_0) being a measured quantity, the desired K_b/K_a would have been determined.

Unfortunately, the setup is not ideal, and hence we will have to write the more complicated relation:

$$\begin{bmatrix} b_0 \\ a_1 \end{bmatrix} = \begin{bmatrix} e_{00} & e_{01} \\ e_{10} & e_{11} \end{bmatrix} \begin{bmatrix} a_0 \\ b_1 \end{bmatrix} \tag{4.14}$$

We have made use of the fact that the system is linear, and consequently, the above two equations are sufficient to cater to all nonidealities discussed earlier. The choice of which two variables to keep on the left and which on the right is quite arbitrary—this particular choice is commonly used probably because in the ideal case (with even the mixer gain, etc., taken into account in the ADC) the matrix will become the unit matrix. The terms $e_{00}, \dots, e_{11}$ are called error terms, although obviously only some are errors in the sense that we would like them to be 0. The equations can also be shown graphically (Figure 4.8).

To interpret such figures, pick any boxed variable, say, a_1. Look at the arrows pointing into this box. There is one arrow coming from the a_0 box, with a label e_{10}, and another from the b_1 box, with label e_{11}. This gives the equation $a_1 = e_{10} a_0 + e_{11} b_1$. Similarly, looking at box b_0, we get the equation $b_0 = e_{00} a_0 + e_{01} b_1$. These two are the equations in (4.16). There is also the additional equation $b_1 = a_1 \Gamma$, which is of course the definition of Γ. Now let us try to express the measured reflection coefficient Γm, which is (b_0/a_0), in terms of Γ and the error terms. Using (4.14) and the relations

$$b_1 = a_1 \Gamma \quad \text{and} \quad \Gamma_m = (b_0/a_0) \tag{4.15}$$

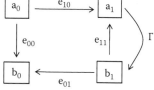

FIGURE 4.8
Signal flow graph for reflection measurement.

we can write after some algebra:

$$\Gamma_m = e_{00} + \frac{e_{01}e_{10}\Gamma}{1 - e_{11}\Gamma} = \frac{e_{00} - \Delta_e\Gamma}{1 - e_{11}\Gamma} \qquad (4.16)$$

Here Δe is used as an abbreviation for $(e_{00}\,e_{11} - e_{10}\,e_{01})$.
    We can also reverse (4.16) and express Γ in terms of Γ_m:

$$\Gamma = \frac{\Gamma_m - e_{00}}{e_{11} - \Gamma_m - \Delta_e} \qquad (4.17)$$

From (4.17) we can see that if the three unknowns—e_{00}, e_{11}, and Δe—are determined, we will be able to obtain the true reflection coefficient Γ from the measured Γ_m.
    This determination is not difficult; rewrite (4.16) as

$$\Gamma_m = e_{00} - \Delta_e\Gamma + e_{11}\Gamma\Gamma_m \qquad (4.18)$$

If three measurements are made with three different known values of Γ, then (4.18) results in three linear simultaneous equations in the three variables: e_{00}, e_{11}, and Δ_e. These can be solved for, and then we can obtain accurate values for any unknown Γ using (4.17). For co-axial measurement systems the three known reflection coefficients are supplied by three calibration standards: the short, the open, and the load. The Γ-values are *close to (but not exactly)* –1, 1, and 0, respectively. For accuracy, the exact values as functions of frequency covering the full frequency range for which the standard is supposed to be used are supplied by the manufacturer. This information is usually called cal-kit. For rectangular metal waveguide systems, the standards used are usually the short, the offset-short, and the load. The short and load are similar to the co-axial case, but the offset-short is a waveguide short with a small length (say, L) of waveguide attached. This L is usually taken as a quarter wavelength at center frequency (so the standard will behave as an open at exactly the center frequency). Let us say that a waveguide is used for the 8 to 12 GHz range, and $L = \lambda g/4$ at 10 GHz. In that case, L will be *roughly* (actually the waveguide is dispersive and the wavelength does not decrease linearly as frequency increases; the student may evaluate the exact numbers using standard expressions for guided wavelength of a waveguide) $0.2\lambda_g$ at 8 GHz and $0.3\lambda_g$ at 12 GHz. So, it is clear that throughout the 8–12 GHz range, the reflection coefficient of the offset-short is substantially different from the short (although both have magnitude = 1). Looking at (4.18), all that is really needed is that the three Γ-values be different so that the equations may be solved—this will be definitely satisfied.

Example 4.2

One port calibration of a 50 Ω VNA is done using an ideal short, an ideal open, and a load that is supposed to be 50 Ω, but is actually 40 Ω. What are the calculated magnitudes of e_{01}, e_{11}, and $e_{01} e_{10}$ if the VNA is ideal (not known to the user) and has $e_{01} = e_{10} = 1$ and $e_{00} = e_{11} = 0$.

We start from (4.18):

$$e_{00} + \Gamma\Gamma m \, e_{11} - \Gamma \, (e_{00} \, e_{11} - e_{01} e_{10}) = \Gamma m \qquad (4.19)$$

Here we have an unusual situation: the VNA is actually ideal, but the user thinks it is not, while the load standard is nonideal while user thinks it is exactly 50 Ω.

The user expects an ideal load, so (4.19) becomes $e_{00} = \Gamma_{m\_Load}$. The VNA is actually ideal, so we can calculate what Γ_{m_Load} will be.

Measured $\Gamma_{m\_Load} = (40-50)/(40+50)$ because VNA is ideal. So, we will conclude that $e_{00} = -1/9$.

For the open standard ($\Gamma m\_open = 1 = \Gamma$) the user will conclude:

$$\frac{-1}{9} + (e_{11}) - \Delta_e = 1$$

And for the short ($\Gamma m\_short = -1 = \Gamma$):

$$\frac{-1}{9} + (e_{11}) + \Delta_e = -1$$

Adding these gives $e_{11} = 1/9$ and $\Delta e = -1$.

4.4.2 SOLT Calibration (Two-Port)

Now we refer to Figure 4.7 and the mathematics becomes much more complicated. The signal-flow graph for the case where the switch is connected to coupler 1 (this is called forward model) is shown in Figure 4.9.

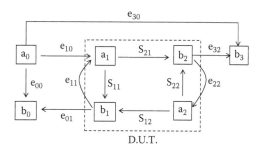

FIGURE 4.9
Signal flow graph for forward model.

Here, a_0, b_0, and b_3 are the digitized signals available from ADC1, ADC2, and ADC3 in Figure 4.6, respectively. The error terms have been traditionally given some names that enable us to associate some physical effects with them:

e_{00}: Directivity. This represents the finite coupler isolation.

e_{11}: Port 1 match. Since the coupler 1 input may not be exactly matched, some reflected signal from DUT port 1 may go back to the DUT from the coupler.

$(e_{10}\, e_{01})$: Reflection tracking. This represents the coupling factor, mixer conversion loss, etc., in the path from the incident signal sent to the coupler to the reflected signal digitized. It should ideally be high.

$(e_{10}\, e_{32})$: Transmission tracking. This represents the coupling factor, mixer conversion loss, etc., in the path from the incident signal sent to the coupler to the transmitted signal digitized. It should ideally be high.

e_{22}: Port 2 match. Similar to e_{11}.

e_{30}: Leakage. This represents the nonideal switch and other effects because of which signal may appear at DUT port 2 without passing through the DUT, while the switch is in the forward path, as shown by the solid arrow in Figure 4.6. For modern instruments it is very low and is usually neglected.

The reason why six terms are defined (using two products) and not seven (although seven terms are shown in Figure 4.8) will be apparent from the equations for the measured S-parameters.

Some of the relevant equations that can be extracted from Figure 4.8 are

$$b_3 = e_{32}\, b_2 + e_{30}\, a_0 \quad \text{and} \quad a_2 = e_{22}\, b_2$$

There are many more equations; combining them, we can arrive at the following expressions for the measured S-parameters $S_{11m} = b_0/a_0$ and $S_{21m} = b_3/a_0$:

$$S_{11m} = e_{00} + \frac{(e_{10}e_{01})(S_{11} - e_{22}\Delta_S)}{1 - e_{11}S_{11} - e_{22}S_{22} + e_{11}e_{22}\Delta_S} \tag{4.20}$$

$$S_{21m} = e_{30} + \frac{(e_{10}e_{32})(S_{21})}{1 - e_{11}S_{11} - e_{22}S_{22} + e_{11}e_{22}\Delta_S} \tag{4.21}$$

where $\Delta_s = S_{11}S_{22} - S_{21}S_{12}$.

It is clear that the six error terms defined above have to be evaluated. This is done in the following way:

Step 1: Connect short, open (or offset-short for waveguide), and load standards to port 1, and measure the three values of S_{11m}. These standards also ensure that $S_{21} = 0$ as there is no connection to port 2. Measured $S_{21}m$ will of course not be 0 but will be e_{30}, which is evaluated right away. When $S_{21} = 0$, $\Delta_s = S_{11}S_{22}$, and we get the simplification

$$1 - e_{11}S_{11} - e_{22}S_{22} + e_{11}e_{22}\Delta_S = 1 - e_{11}S_{11} - e_{22}S_{22} + e_{11}e_{22}S_{11}S_{22}$$

$$= (1 - e_{11}S_{11})(1 - e_{22}S_{22})$$

This gives

$$S_{11m} = e_{00} + \frac{(e_{10}e_{01})(S_{11} - e_{22}S_{11}S_{22})}{1 - e_{11}S_{11} - e_{22}S_{22} + e_{11}e_{22}\Delta_S} = e_{00} + \frac{(e_{10}e_{01})(S_{11})}{1 - e_{11}S_{11}} \quad (4.22)$$

This is exactly the same as (4.16), and hence e_{00}, e_{11}, and $\Delta_e = (e_{00}\,e_{11} - e_{01}e_{10})$ can be evaluated in the same way. The reflection tracking term $(e_{10}\,e_{01}) = e_{00}\,e_{11} - \Delta_e$, and hence is known.

Step 2: Connect ports 1 and 2 directly, omitting any DUT. This is called a through connection and has S-parameters: $S_{11} = S_{22} = 0$ and $S_{12} = S_{21} = 1$. Now (4.20) becomes

$$S_{11m}e_{00} + \frac{(e_{10}e_{01})e_{22}}{1 - e_{11}e_{22}} = \frac{e_{00} - e_{22}\Delta_e}{1 - e_{11}e_{22}} \quad (4.23)$$

Rearranging this we get

$$e_{22} = \frac{S_{11m} - e_{00}}{S_{11m}e_{11} - \Delta_e},$$

from which e_{22} can be evaluated.

Also (4.21) becomes

$$S_{21m} = e_{30} + \frac{(e_{10}e_{32})}{1 - e_{11}e_{22}} \quad (4.24)$$

Now the transmission tracking $(e_{10}\,e_{32}) = (S_{21m} - e_{30})(1 - e_{11}\,e_{22})$. Since all terms on the right-hand side (rhs) are known at this stage, $e_{10}\,e_{32}$ is determined, and hence the six error terms defined until now are known.

Let us now consider the situation when the switch position in Figure 4.7 is changed—this is called the reverse model. Now the signal-flow graph is shown in Figure 4.10.

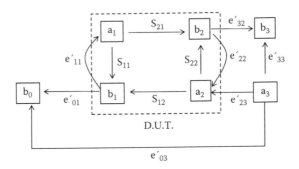

FIGURE 4.10
Signal flow graph for reverse model.

In this case, a_3 represents the ADC1 output, b_0 the ADC2 output, and b_3 the ADC3 output. The new error terms are shown with a prime to distinguish these from the earlier ones. They are:

e'_{33}: Directivity. This represents the finite coupler isolation.

e'_{11}: Port 1 match. Since the coupler 2 input may not be exactly matched, some reflected signal from DUT port 1 may go back to the DUT from the coupler.

$(e'_{23}\ e'_{32})$: Reflection tracking. This represents the coupling factor, mixer conversion loss, etc., in the path from the incident signal sent to coupler 2 to the reflected signal, digitized by ADC3. It should ideally be high.

$(e'_{23}\ e'_{01})$: Transmission tracking. This represents the coupling factor, mixer conversion loss, etc., in the path from the incident signal sent to coupler 2 to the transmitted signal, digitized by ADC2. It should ideally be high.

e'_{22}: Port 2 match. Similar to e'_{11}. Since the switch change means that the circuit has changed, we do not assume that $e'_{22} = e_{22}$, etc.

e'_{03}: Leakage. This represents the nonideal switch and other effects because of which signal may appear at DUT port 1 without passing through the DUT, while the switch is in the reverse path, as shown by the dashed arrow in Figure 4.6.

Now the counterparts of Equations (4.20) and (4.21) are

$$S_{22m} = e'_{33} + \frac{(e'_{23}e'_{32})(S_{22} - e'_{11}\Delta_S)}{1 - e'_{11}S_{11} - e'_{22}S_{22} + e'_{11}e'_{22}\Delta_S} \tag{4.25}$$

$$S_{12m} = e'_{03} + \frac{(e'_{23}e'_{01})(S_{12})}{1 - e'_{11}S_{11} - e'_{22}S_{22} + e'_{11}e'_{22}\Delta_S} \tag{4.26}$$

These are identical in form to (4.20) and (4.21), and hence the same steps can be adopted.

Notice that with short/open/load connected to port 2, once again S_{12} is 0 and S_{12m} directly gives e'_{03}. If $S_{21} = 0$, then as before, S_{22m} becomes

$$S_{22m} = e'_{33} + \frac{(e'_{23}e'_{32})(S_{22})}{1 - e'_{22}S_{22}} \tag{4.27}$$

Using the same on-port calibration method now at port 2, we can evaluate e'_{33}, e'_{22}, and $(e'_{23}\, e'_{32})$.

With the through connection as before, (4.25) reduces to

$$e'_{11} = \frac{S_{22m} - e'_{33}}{S_{22m}e'_{22} - \Delta'_e}, \text{ where } \Delta'_e = e'_{33}e'_{22} - e'_{32}e'_{23} \tag{4.28}$$

So, e'_{11} is evaluated. Also with the through connection,

$$S_{12m} = e'_{03} + \frac{(e'_{23}e'_{01})}{1 - e'_{11}e'_{22}} \tag{4.29}$$

Since e'_{03} has already been evaluated above, the last unknown $e'_{23}\, e'_{01}$ is evaluated as

$$(e'_{23}\, e'_{01}) = (S_{12}m - e'_{03})(1 - e'_{11}e'_{22})$$

Now all 12 error terms have been evaluated, and the true S-parameters can be evaluated from any subsequent measured S-parameters using the following equations (obtained by solving Equations (4.20), (4.21), (4.25), and (4.26) for S_{11}, S_{12}, S_{21}, and S_{22}):

$$S_{11} = \left[\frac{S_{11m} - e_{00}}{e_{10}e_{01}}\left(1 + \frac{S_{22m} - e'_{33}}{e'_{23}e'_{32}}e'_{22} \right) - e_{22}\left(\frac{S_{21m} - e_{30}}{e_{10}e_{32}} \right)\left(\frac{S_{12m} - e'_{03}}{e'_{23}e'_{01}} \right) \right]\Big/ D \tag{4.30a}$$

$$S_{22} = \left[\frac{S_{22m} - e'_{33}}{e'_{23}e'_{32}}\left(1 + \frac{S_{11m} - e_{00}}{e_{10}e_{01}}e_{11} \right) - e'_{11}\left(\frac{S_{21m} - e_{30}}{e_{10}e_{32}} \right)\left(\frac{S_{12m} - e'_{03}}{e'_{23}e'_{01}} \right) \right]\Big/ D \tag{4.30b}$$

$$S_{21} = \left(\frac{S_{21m} - e_{30}}{e_{10}e_{32}} \right)\left(1 + \frac{S_{22m} - e'_{33}}{e'_{23}e'_{32}}(e'_{22} - e_{22}) \right)\Big/ D \tag{4.30c}$$

$$S_{12} = \left(\frac{S_{21m} - e'_{03}}{e'_{23}e'_{01}} \right)\left(1 + \frac{S_{11m} - e_{00}}{e_{10}e_{01}}(e_{11} - e'_{11}) \right)\Big/ D \tag{4.30d}$$

Here,

$$D=\left(1+\frac{S_{11m}-e_{00}}{e_{10}e_{01}}e_{11}\right)\left(1+\frac{S_{22m}-e'_{33}}{e'_{23}e'_{32}}e'_{22}\right)-\left(\frac{S_{21m}-e_{30}}{e_{10}e_{32}}\right)\left(\frac{S_{12m}-e'_{03}}{e'_{23}e'_{01}}\right)(e_{22}e'_{11})$$

(4.30e)

It can be seen that we have not assumed that the short/open/load standards were ideal but have taken ideal S-parameters for the through standard. This is actually not an assumption. Let us take a closer look at the connection between the VNA cable and the DUT shown as a black rectangle Figure 4.10.

The VNA – cables are shown as thick black lines and the DUT is shown as a black rectangle. In between is the DUT connector—is this part of the VNA or part of the DUT? As we can see from the zoomed view of the connector, it is not clear exactly where the VNA ends and the DUT begins. Since this ambiguity is several mm in length (the length of a typical connector), it will have a significant effect on the phase, and cannot be neglected. One benefit of calibration is that this ambiguity should be resolved.

Figure 4.12 shows the connector from Figure 4.11 for the four cases: short, open, load, and through connections. In Figure 4.12, focus on the short standard. Suppose that the data specifying the reflection coefficient of this (close to –1) are given with respect to plane A. In that case, it is essential

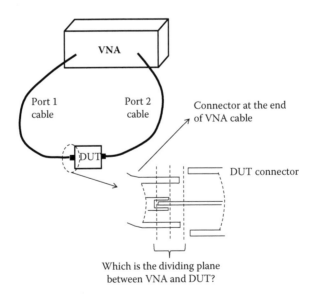

FIGURE 4.11
Identifying the junction of the VNA and the out.

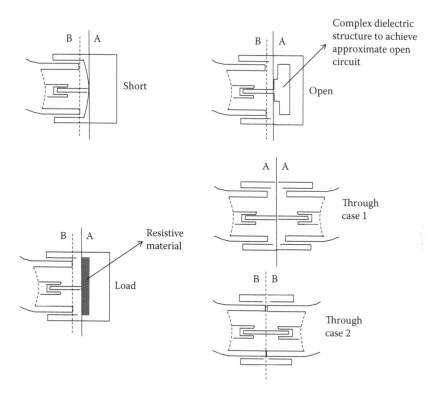

FIGURE 4.12
Zoomed view of the calibration process.

that the data for the open and load standards are also specified with respect to exactly this plane A. For the open standard in particular, it may indeed turn out that the reflection coefficient is closest to +1 when the reference plane is some other plane, B, for example. But if we have chosen plane A for the short, then we obviously have to stick to it. Now consider the through (case 1). Physically, this is implemented with a short co-axial connector of a precise length—notice that the distance from the VNA cable end to plane A has to maintained in all four standards. The length of the connector for the through (case 1) is calculated from this consideration. Now there is no confusion in the dividing plane between the VNA and the DUT—it is plane A in Figure 4.11 (through, case 1). It is very important that whatever is to the left of this dividing plane is left exactly as it is in Figure 4.12 (notice that this is maintained in all four cases—short, open, load, and through case 1) for any subsequent measurement.

Alternatively, we could also have carried out the entire process with plane B. The short standard measured with this reference may show a reflection coefficient that is slightly different from –1, but as long as the correct values are used for all four cases, the calibration will work. Looking at the configuration of through case 2, the dividing plane is now exactly at the tip of the VNA cable in Figure 4.12.

Finally, consider the through standard (case 1 or case 2). It is what is connected between the dividing planes as we approach the device from the VNA cable, from the left or right. But the dividing planes have merged to the single plane A. So, the through is actually a zero-length co-axial line—obviously this has S-parameters of $S_{11} = S_{22} = 0$ and $S_{21} = S_{12} = 1$. Sometimes the through standard cannot be designed to be exactly the desired length. To cater to this and other problems, an extended version of the SOLT calibration, the SOLR (R standing for *reciprocal*), has been developed, where the exact properties of the through standard are not required, as long as it is reciprocal.

4.4.3 TRL Calibration

It was seen that the SOLT calibration is suitable for co-axial or waveguide measurements. But most circuits are built using planar transmission lines—microstrip in particular. Fabricating short, open, and load standards in microstrip is difficult, and even if this is done, characterizing them is a major problem. It will obviously be desirable to have a calibration technique that is more general—not restricted to any particular type of transmission line. The TRL calibration (TRL stands for through-reflect-line) satisfies this requirement; it was first described in Engen and Hoer (1992). Obviously the one standard that can definitely be realized in any medium is the line standard— a certain length of transmission line. If the line length is reduced to 0, we get the through standard. The reflect standard should provide a high reflection, but the *exact value of the reflection coefficient is not required*. An example in microstrip is shown in Figure 4.13.

The line standard in this case is a length L of microstrip. The through standard by definition is of length = 0, but some length d of microstrip is kept between the co-axial connector and the reference plane. The final connection to the VNA will be through co-axial cables. This length d implies that for actual measurement, the reference planes are this same d away from the co-axial connectors. For actual measurements, the circuit has to be kept within these reference planes, as shown in Figure 4.13—the co-axial connectors should not be closer than d from the ends of the circuit that is to be characterized.

The model for the TRL calibration is slightly different. Normally an 8-term model is used—this reduction from the 12-term model for the previous case results from ignoring the isolation terms e_{30} and e'_{03}. The model is now easily represented using two-port S-parameters for the error terms, as in Figure 4.14.

The S-parameters of the error boxes are the unknowns that are to be determined (in an ideal case these would take the form $S_{11} = S_{22} = 0$ and $S_{12} =$

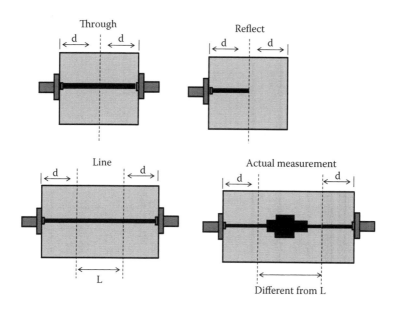

FIGURE 4.13
TRL calibration in microstrip.

$S_{21} = 1$, in which case the measured S-parameters would be the same as the DUT S-parameters). Let us see how the calibration is done.

We will use T-parameters rather than S-parameters here. The T-parameters have the property that overall $[T]$ of a cascade is the product of the individual $[T]$-matrices. Of course, $[S]$ and $[T]$ are equivalent and the conversion from one to the other is given below.

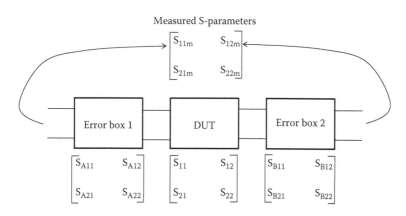

FIGURE 4.14
TRL calibration model.

T-parameters are defined by

$$
\begin{bmatrix} b_1 \\ a_1 \end{bmatrix} = \begin{bmatrix} T_{11} & T_{12} \\ T_{21} & T_{22} \end{bmatrix} \begin{bmatrix} a_2 \\ b_2 \end{bmatrix}
$$

Recall the definition of *S*-parameters:

$$
\begin{bmatrix} b_1 \\ b_2 \end{bmatrix} = \begin{bmatrix} S_{11} & S_{12} \\ S_{21} & S_{22} \end{bmatrix} \begin{bmatrix} a_1 \\ a_2 \end{bmatrix}
$$

Conversion formulae are:

$$
\begin{bmatrix} T_{11} & T_{12} \\ T_{21} & T_{22} \end{bmatrix} = \begin{bmatrix} -\left(\dfrac{S_{11}S_{22}-S_{12}S_{21}}{S_{21}}\right) & \dfrac{S_{11}}{S_{21}} \\ -\dfrac{S_{22}}{S_{21}} & -\dfrac{1}{S_{21}} \end{bmatrix}
$$

$$
\begin{bmatrix} S_{11} & S_{12} \\ S_{21} & S_{22} \end{bmatrix} = \begin{bmatrix} \dfrac{T_{12}}{T_{22}} & \dfrac{T_{11}T_{22}-T_{12}T_{21}}{T_{22}} \\ \dfrac{1}{T_{22}} & -\dfrac{T_{21}}{T_{22}} \end{bmatrix}
$$

Let us consider the TRL standards through and line. *S*- and *T*-parameters of the through are

$$
[S_T] = \begin{bmatrix} 0 & 1 \\ 1 & 0 \end{bmatrix} \qquad [T_T] = \begin{bmatrix} 1 & 0 \\ 0 & 1 \end{bmatrix}
$$

S- and *T*-parameters of the line are

$$
[S_L] = \begin{bmatrix} 0 & e^{-\delta l} \\ e^{-\delta l} & 0 \end{bmatrix} \qquad [T_L] = \begin{bmatrix} e^{-\delta l} & 0 \\ 0 & e^{\delta l} \end{bmatrix}
$$

Here, δ is the complex propagation constant of the line. The more common symbol is γ, but it will be used for a different quantity here.

Using the cascading property of *T*-parameters,

$$
[T_M] = [T_A][T_{DUT}][T_B] \tag{4.31}
$$

where $[T_M]$ is overall measured, $[T_A]$ and $[T_B]$ are those of error boxes A and B, respectively, $[T_{DUT}]$ and is that of DUT:

$$[T_{DUT}] = [T_A]^{-1}[T_M][T_B]^{-1} \tag{4.32}$$

To determine $[T_A]$ and $[T_B]$ from measurements, initially it looks like eight unknown terms, but actually seven are sufficient.

Let

$$[T_A] = p \begin{bmatrix} a & b \\ c & 1 \end{bmatrix} \qquad [T_B] = q \begin{bmatrix} \alpha & \beta \\ \gamma & 1 \end{bmatrix} \tag{4.33}$$

Taking the inverses for (4.32),

$$[T_A]^{-1} = \frac{1}{p(a-bc)} \begin{bmatrix} 1 & -b \\ -c & a \end{bmatrix}$$

$$[T_B]^{-1} = \frac{1}{q(\alpha-\beta\gamma)} \begin{bmatrix} 1 & -\beta \\ -\gamma & \alpha \end{bmatrix}$$

Hence, (4.32) becomes

$$[T_{DUT}] = \frac{1}{pq(a-bc)(\alpha-\beta\gamma)} \begin{bmatrix} 1 & -b \\ -c & a \end{bmatrix}[T_M]\begin{bmatrix} 1 & -\beta \\ -\gamma & \alpha \end{bmatrix} \tag{4.34}$$

Thus, *seven* unknowns are sufficient: a, b, c, α, β, γ, and (pq).

Step 1: Use the through and line standards. If we examine (4.31), we see that T_A and T_B are combined. Let us try to separate them using some matrix algebra.

Let $[T_{M\_T}]$ be the measured $[T]$ using through standard

$$[T_{M\_T}] = [T_A][T_T][T_B]$$

$$[T_{M\_T}] = [T_A]\begin{bmatrix} 1 & 0 \\ 0 & 1 \end{bmatrix}[T_B] \tag{4.35}$$

$$[T_{M\_T}] = [T_A][T_B]$$

Let $[T_{M\_T}]$ be the measured $[T]$ from line standard

$$[T_{M\_L}] = [T_A][T_L][T_B] \tag{4.36}$$

Let us try to remove $[T_B]$. From (4.35),

$$[T_B] = [T_A]^{-1}[T_{M\_T}] \qquad (4.37)$$

Substituting (4.36) in (4.37),

$$[T_{M\_L}] = [T_A][T_L][T_A]^{-1}[T_{M\_T}]$$

$$\Rightarrow [T_{M\_L}][T_{M\_T}]^{-1}[T_A] = [T_A][T_L]$$

$$\Rightarrow [T_{M\_Comb}][T_A] = [T_A][T_L] \qquad (4.38)$$

where $[T_{M\_Comb}] = [T_{M\_L}][T_{M\_T}]^{-1} = \begin{bmatrix} t_{11} & t_{12} \\ t_{21} & t_{22} \end{bmatrix}$ is known data

obtained from measurement of standards.

On expanding the matrices (4.38) becomes

$$\begin{bmatrix} t_{11} & t_{12} \\ t_{21} & t_{22} \end{bmatrix} p \begin{bmatrix} a & b \\ c & 1 \end{bmatrix} = p \begin{bmatrix} a & b \\ c & 1 \end{bmatrix} \begin{bmatrix} e^{-\delta l} & 0 \\ 0 & e^{\delta l} \end{bmatrix} \qquad (4.39)$$

Expanding this gives the following four scalar equations:

$$t_{11}a + t_{12}c = a\, e^{-\delta l} \qquad (4.40a)$$

$$t_{11}b + t_{12} = b\, e^{\delta l} \qquad (4.40b)$$

$$t_{21}a + t_{22}c = c\, e^{-\delta l} \qquad (4.40c)$$

$$t_{21}b + t_{22} = e^{\delta l} \qquad (4.40d)$$

Dividing (4.40a) by (4.40c) gives

$$\frac{t_{11}a + t_{12}c}{t_{21}a + t_{22}c} = \frac{a}{c}$$

$$\Rightarrow \frac{t_{11}\dfrac{a}{c} + t_{12}}{t_{21}\dfrac{a}{c} + t_{22}} = \frac{a}{c}$$

On rearranging we get a quadratic equation in $\frac{a}{c}$:

$$t_{21}\left(\frac{a}{c}\right)^2 + (t_{22} - t_{11})\left(\frac{a}{c}\right) - t_{12} = 0 \tag{4.41}$$

Similarly, (4.40b) and (4.40d) give

$$\frac{t_{11}b + t_{12}}{t_{21}b + t_{22}} = b$$

$$\Rightarrow t_{21}(b)^2 + (t_{22} - t_{11})(b) - t_{12} = 0 \tag{4.42}$$

Remember that $t_{11}, t_{21}, t_{12}, t_{22}$, and are all known numbers from measurement of through and line standards. Hence, they are known coefficients of (4.41) and (4.42).

Since (4.41) and (4.42) have the same coefficients, hence $\frac{a}{c}$ and b have to be two roots of the quadratic equation. This is a good start. However, which is which is yet to be determined.

For that, let us see how a, b, and c correspond to the S-parameters of error box A. (Remember, a, b, and c are defined in T-parameters of error box A.)

$$[T_A] = p \begin{bmatrix} a & b \\ c & 1 \end{bmatrix}$$

$$[S_A] = \begin{bmatrix} s_{A11} & s_{A12} \\ s_{A21} & s_{A22} \end{bmatrix}$$

Using the T-to-S conversion table repeated here, we get the relation of a, b, c, and p in terms of $s_{A11}, s_{A21}, s_{A12}$, and s_{A22}:

$$\begin{bmatrix} T_{11} & T_{12} \\ T_{21} & T_{22} \end{bmatrix} = \begin{bmatrix} -\left(\dfrac{S_{11}S_{22} - S_{12}S_{21}}{S_{21}}\right) & \dfrac{S_{11}}{S_{21}} \\ -\dfrac{S_{22}}{S_{21}} & \dfrac{1}{S_{21}} \end{bmatrix}$$

So,

$$p = \frac{1}{s_{A21}}, \; = s_{A11}, \; c = -s_{A22}, \text{ and } a = -(s_{A11}s_{A22} - s_{A21}s_{A12})$$

$$\Rightarrow \frac{a}{c} = s_{A11} - \frac{s_{A21}s_{A12}}{s_{A22}}$$

In most cases where the error box mismatch is small, then $|s_{A11}| \ll 1$, (so is $|s_{A22}|$) and $|s_{A21}| \approx 1$ (so is $|s_{A12}|$).

Therefore, we conclude that

$$\left|\frac{a}{c}\right| \gg |b|$$

Thus, a root with a larger magnitude is (a/c) and with a smaller magnitude is (b). Hence, a/c and (b) are now completely determined from the quadratic equation (4.41) or (4.42), which are identical.

Additionally, dividing (4.40c) by (4.40d) gives

$$\frac{t_{21}a + t_{22}c}{t_{21}b + t_{22}} = c\, e^{-2\delta l}$$

$$\Rightarrow \frac{t_{21}\left(\dfrac{a}{c}\right) + t_{22}}{t_{21}b + t_{22}} = e^{-2\delta l} \tag{4.43}$$

Note that since we roughly know δl (so the phase of the rhs is roughly known), (4.43) can also be used to choose the roots correctly—the wrong root choice will reverse the phase of the expression on the left-hand side (lhs).

Equation (4.43) can be used to determine the propagation constant of the transmission line for the substrate, and thus helps to characterize the substrate during calibration.

Step 2: Now that we have determined some of the unknowns in error box A (a/c and b), let us see if can we determine some unknowns of error box B. From (4.37), repeated here:

$$[T_B] = [T_A]^{-1}[T_{M\_T}]$$

Remember that $[T_{M\_T}]$ is known from measurement of the through standard.

Let $[T_{M\_T}]$ be expressed as (note that d, e, f, and r are all known numbers):

$$[T_{M\_T}] = r\begin{bmatrix} d & e \\ f & 1 \end{bmatrix}$$

Hence, (4.37) becomes

$$q\begin{bmatrix} \alpha & \beta \\ \gamma & 1 \end{bmatrix} = \frac{1}{p(a-bc)}\begin{bmatrix} 1 & -b \\ -c & a \end{bmatrix}r\begin{bmatrix} d & e \\ f & 1 \end{bmatrix}$$

$$\Rightarrow \begin{bmatrix} \alpha & \beta \\ \gamma & 1 \end{bmatrix} = \frac{r}{pq(a-bc)}\begin{bmatrix} d-bf & e-b \\ af-cd & a-ce \end{bmatrix}$$

Thus, we get

$$pq = \frac{r(a-ce)}{(a-bc)} \Rightarrow pq = \frac{r\left(\dfrac{a}{c}-e\right)}{\left(\dfrac{a}{c}-b\right)},$$

so pq is determined.

$$\alpha = \frac{(d-bf)}{(a-ce)} \Rightarrow a\alpha = \frac{(d-bf)}{\left(1-\dfrac{c}{a}e\right)} \tag{4.44a}$$

$$\beta = \frac{(e-b)}{(a-ce)} \tag{4.44b}$$

$$\gamma = \frac{(af-cd)}{(a-ce)} \Rightarrow \gamma = \frac{\left(\dfrac{a}{c}f-d\right)}{\left(\dfrac{a}{c}-e\right)}$$

and γ is also found.

Thus, pq and γ are completely determined. α and β require a to be found first.

Step 3: Thus, if a is determined, then all the required values would be determined. Now let us try to see if can we determine a from the reflect standard. Reflect standard at port 1 and measurement of S_{11} (call this measured value SM_R_1):

$$S_{M\_R1} = s_{A11} + \frac{s_{A21}s_{A12}\Gamma_r}{1-s_{A22}\Gamma_r}$$

The same reflect standard at port 2 and measurement of S_{22} (call this measured value S_{M_R1}):

$$S_{M\_R2} = s_{B22} + \frac{s_{B21}s_{B12}\Gamma_r}{1 - s_{B11}\Gamma_r}$$

Here, we have used standard results for the reflection coefficient looking into one port of a black box when the other port is terminated with some given reflection coefficient (Ha, 1981). The student may also try to derive this simple relation using basic S-parameter relation discussed in Chapter 2.

Replacing S-parameters by T-parameters:

$$S_{M\_R1} = S_{A11} + \frac{S_{A21}S_{A12}\Gamma_r}{1 - S_{A22}\Gamma_r}$$

$$\Rightarrow S_{M\_R1} = \frac{T_{A12}}{T_{A22}} + \frac{\left(\dfrac{1}{T_{A22}}\right)\left(\dfrac{T_{A11}T_{A22} - T_{A21}T_{A12}}{T_{A22}}\right)\Gamma_r}{1 - \left(-\dfrac{T_{A21}}{T_{A22}}\right)\Gamma_r}$$

$$\Rightarrow S_{M\_R1} = \frac{T_{A12} + T_{A11}\Gamma_r}{T_{A22} + T_{A21}\Gamma_r}$$

But $T_{A11} = pa$, $T_{A12} = pb$, $T_{A21} = pc$, and $T_{A22} = p$
Hence, we get

$$S_{M\_R1} = \frac{b + a\Gamma_r}{1 + c\Gamma_r}$$

Or, solving for Γ_r,

$$\Gamma_r = \frac{S_{M\_R1} - b}{a - cS_{M\_R1}} \tag{4.45}$$

Hence,

$$a = \frac{S_{M\_R1} - b}{\left(1 - \dfrac{c}{a}S_{M\_R1}\right)\Gamma_r} \tag{4.46}$$

Similarly, processing SM_R_2:

$$S_{M\_R2} = s_{B22} + \frac{s_{B21}s_{B12}\Gamma_r}{1 - s_{B11}\Gamma_r}$$

$$S_{M\_R2} = -\frac{T_{B21}}{T_{B22}} + \frac{\left(\dfrac{1}{T_{B22}}\right)\left(\dfrac{T_{B11}T_{B22} - T_{B21}T_{B12}}{T_{B22}}\right)\Gamma_r}{1 - \left(\dfrac{T_{B12}}{T_{B22}}\right)\Gamma_r}$$

$$S_{M\_R2} = \frac{T_{B11}\Gamma_r - T_{B21}}{T_{B22} - T_{B12}\Gamma_r}$$

$$S_{M\_R2} = \frac{\alpha\Gamma_r - \gamma}{1 - \beta\Gamma_r}$$

Solving for Γ_r:

$$\Gamma_r = \frac{S_{M\_R2} + \gamma}{\alpha + \beta S_{M\_R2}}$$

Hence,

$$\alpha = \frac{S_{M\_R2} + \gamma}{\left(1 + \dfrac{\beta}{\alpha} S_{M\_R2}\right)\Gamma_r} \tag{4.47}$$

Note that $\frac{\beta}{\alpha}$ is completely known, since dividing (4.44b) by (4.44a):

$$\frac{\beta}{\alpha} = \frac{e - b}{d - bf}$$

Now dividing (4.46) by (4.47) gives

$$\frac{a}{\alpha} = \frac{(S_{M\_R1} - b)\left(1 + \dfrac{\beta}{\alpha} S_{M\_R2}\right)}{(S_{M\_R2} + \gamma)\left(1 - \dfrac{c}{a} S_{M\_R1}\right)} \tag{4.48}$$

Multiplying (4.44a) and (4.48) we get

$$a^2 = \frac{(d-bf)(S_{M\_R1}-b)\left(1+\dfrac{\beta}{\alpha}\,S_{M\_R2}\right)}{\left(1-\dfrac{c}{a}e\right)(S_{M\_R2}+\gamma)\left(1-\dfrac{c}{a}S_{M\_R1}\right)} \tag{4.49}$$

Thus, a can be determined. But which root to take?

From (4.45) calculate the angle of $\frac{S_{M\_R1}-b}{a-cS_{M\_R1}}$ for both roots of a obtained from (4.49). Then select that value of a that gives the above angle closer to the known reflect standard (i.e., $0°$ for open and $180°$ for short).

Thus, a is evaluated without ambiguity. Using (4.44a), α can be evaluated, and using (4.44b), β can be evaluated. Also, (a/c) was determined in (4.41) and (4.43), which means now c is also determined.

Now all the required values are determined and (4.34) can be used to determine the calibrated S-parameter of DUT.

The procedure can be summarized in the following way:

Determining the error terms:

1. Measure S-parameters of through $[SM\_T]$, line $[SM\_L]$, and reflect at both ports SM_R_1 and SM_R_2. Also to be known is the type of reflect standard (i.e., either short or open).
2. Convert S-parameter to T-parameter, $[S_{M\_T}]$ to $[TM\_T]$ and $[SM\_L]$ to $[TM\_L]$.
3. Step 1: Determination of a/c and b:

    a. $[T_{MComb}] = [T_{M\_L}][T_{M\_T}]^{-1} = \begin{bmatrix} t_{11} & t_{12} \\ t_{21} & t_{22} \end{bmatrix}$

    b. Calculate roots of: $[t_{21} \quad x^2 + (t_{22}-t_{11})\; x \quad -t_{21} = 0]$

    c. Larger magnitude root is $\frac{a}{c}$

    c. Smaller magnitude root is b

    d. Or, the choice that gives the correct phase in

$$\frac{t_{21}\left(\dfrac{a}{c}\right)+t_{22}}{t_{21}b+t_{22}} = e^{-2\delta l}$$

This can be used to determine the substrate properties if required.

4. Step 2: Determination of γ, β/α, and pq:
   a. $[T_{M\_T}]$ is Normalized to determine r, d, e, and f
   b.

$$\gamma = \frac{\left(\dfrac{a}{c}f - d\right)}{\left(\dfrac{a}{c} - e\right)}$$

   c.

$$pq = \frac{r\left(\dfrac{a}{c} - e\right)}{\left(\dfrac{a}{c} - b\right)}$$

   d.

$$\frac{\beta}{\alpha} = \frac{e - b}{d - bf}$$

5. Step 3: Determination of a:
   a.

$$a^2 = \frac{(d - bf)(S_{MR1} - b)\left(1 + \dfrac{\beta}{\alpha}S_{MR2}\right)}{\left(1 - \dfrac{c}{a}e\right)(S_{MR2} + \gamma)\left(1 - \dfrac{c}{a}S_{MR1}\right)}$$

   b. Find the two roots, say, a_1 and a_2.
   c. Calculate angle of $\Gamma_r = \frac{S_{MR1} - b}{a - cS_{M\_R1}}$ for both $a1$ and $a2$.
   d. Choose that value that gives an angle close to the known reflect angle.
   e. Then find

$$\alpha = \frac{(d - bf)}{(a - ce)}$$

   f.

$$\beta = \frac{(e - b)}{(a - ce)}$$

6. Verification:

    a. The calibrated *T*-matrix of the through:

$$[T_{C\_T}] = \frac{1}{pq(a-bc)(\alpha-\beta\gamma)}\begin{bmatrix} 1 & -b \\ -c & a \end{bmatrix}[T_{M\_T}]\begin{bmatrix} 1 & -\beta \\ -\gamma & \alpha \end{bmatrix}$$

    b. Convert to *S* parameter $[S_{C\_T}]$, and ideally this should be the identity matrix.

Characterizing DUT after calibration:

1. Two-port DUT:

    a. Measure the *S*-parameters of DUT: $[SM\_DUT]$.

    b. $[SM\_DUT] \rightarrow [TM\_DUT]$

    c.

$$[T_{C\_DUT}] = \frac{1}{pq(a-bc)(\alpha-\beta\gamma)}\begin{bmatrix} 1 & -b \\ -c & a \end{bmatrix}[T_{M\_DUT}]\begin{bmatrix} 1 & -\beta \\ -\gamma & \alpha \end{bmatrix}$$

    d. $[TC\_DUT] \rightarrow [SC\_DUT]$

    $[SC\_DUT]$ is the required calibrated *S*-parameters of two-port DUT.

2. One-port DUT:

    a. Measure the *S*-parameter of DUT at port 1: SM_1_DUT.

    b. $S_{C1\_DUT} = \frac{SM1\_DUT-b}{a-cSM1\_DUT}$ gives the calibrated S_{11}.

Limitation of TRL:

- When $e^{-\delta l} = \pm 1$, then (a/c) becomes equal to b.
- Thus, $a - bc = 0$. Thus, $[T_A]^{-1}$ does not exist, and hence step 2 cannot be carried out. So, α, β, γ, and pq cannot be determined. So, l should not be a multiple of a half wavelength.
- $e^{-\delta l} = \pm j$ is the best choice, of which $-j$ gives the maximum bandwidth (quarter wavelength).
- Thus, TRL is band limited. If l is a quarter wavelength at some f_0, then it can be used up to a maximum that is well below $2f_0$. Also, below $f_0/4$, the electrical length of the through standard becomes $22.5°$, and it is not advisable to go much below this to distinguish the line and the through.
- For broadband applications, multiple line standards at different bands are required.

4.5 Frequency Offset and Mixer Measurement

While originally the VNA was developed for characterizing linear time-invariant components in terms of S-parameters, today this instrument is used for a wide variety of nonlinear measurements. The simplest of these is the mixer measurement. For this the mixer is regarded as a linear component from the radio frequency (RF) to the IF port, and the LO is assumed to be an internal part of the mixer. Usually, a network analyzer intended for mixer measurement will have an in-built extra synthesized source to supply the LO, although it can also be taken from an external source. The modified VNA is shown in Figure 4.15.

There are different ways in which mixer measurements can be done. In the example shown in Figure 4.14, the VNA itself supplies the LO. Suppose the mixer is designed for an RF range of 8–12 GHz, LO of 8.1–12.1 GHz, and a constant IF of 0.1 GHz. Let us say that the VNA works with an IF of 1 MHz.

To test this, first the microwave source A will sweep from 8 to 12 GHz, while source B sweeps from 7.999 to 11.999 GHz. This will enable ADC1 to record the RF power fed to the mixer, and at the same time the RF power reflected from the mixer input (i.e., the input match of the mixer) is recorded by ADC2. The LO provided to the mixer from the VNA would at the same time sweep from 8.1 to 12.1 GHz. The reading of ADC3 is not used.

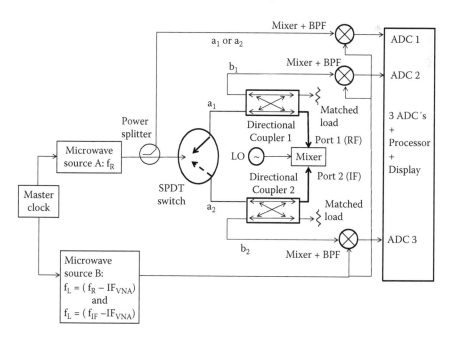

FIGURE 4.15
VNA configured for mixer measurement.

Next, the same sweep is repeated where source A and the internal LO are concerned, but microwave source B is now kept constant at 0.099 GHz. This is 1 MHz below the IF of the mixer (not to be confused with the IF of the VNA). ADC3 now reads the output of the mixer, and conversion gain can be calculated. Phase readings are usually not required, but some interpretations can be made (Dunsmore, 2012).

It is clear that once we allow independent control of the three sources (sources A and B and also the internal LO), many more types of frequency sweeps can be carried out. Modern network analyzers also contain digitally controlled attenuators just before the sources (A, B, and internal LO) reach the external connectors on the VNA. Using this, the swept variable can even be power rather than frequency.

Many more measurements are described in Dunsmore (2012).

4.6 Time Gating

Suppose we are carrying out the scattering measurement using a horn antenna in an antenna chamber shown in Figure 4.16.

Our interest is in sensing the signal reflected from the ball, but there will also be a significant signal reflected from the wall behind it. Although the absorbers minimize this reflection, it cannot be neglected since the area of the absorbing wall is large.

We can make use of the physical separation between the ball and the wall behind it. Suppose the distance from the antenna to the ball is d_b, while the distance from the ball to the wall behind it is d_w. Now suppose we consider the inverse Fourier transform (IFT) of $S_{11}(f)$. This gives us the impulse response. We can guess that it will take the rough shape shown in Figure 4.16.

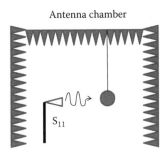

Antenna chamber

FIGURE 4.16
Scattering measurement through S_{11}.

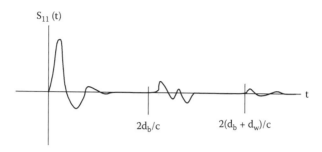

FIGURE 4.17
Typical impulse response for the scattering measurement.

The first large peak is caused by reflection at the horn itself (horn input impedance is not 50 Ω), the next signals are from the ball, and the last peaks are from the wall. If we could see the time domain data, we could actually verify that these peaks occur roughly at the time values shown. Now we want to get rid of the effect of the reflection at the antenna and the wall. This is not possible in the frequency domain, as these all occupy the same band (the horn bandwidth), but in the time domain it is easy. Simply set the values of $S_{11}(t)$ to 0 for $t < 2d_b/c$ and $t > 2(d_b + d_w)/c$. This is the same as multiplying the impulse response by a unit rectangular pulse from $2d_b/c$ to $2(d_b + d_w)/c$. These are, of course, the round-trip times at light speed from the antenna to the ball and from the antenna to the wall. This will give us the gated impulse response, as in Figure 4.18.

Now if we take the Fourier transform, we will get the S_{11}, which will result if the only response was from the ball. This information can be used for radar cross section and other calculations.

In practice, the implementation is slightly different. Since multiplication by a rectangular pulse in the time domain corresponds to convolution with a

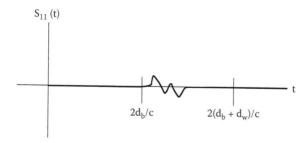

FIGURE 4.18
Gated impulse response.

sinc function in the frequency domain, this is what is done. More precisely, the Fourier transform of a unit pulse from $2d_b/c$ to $2(d_b + d_w)/c$ is

$$e^{-\frac{j2\pi f(2d_b+d_w)}{c}} sinc\left(f\frac{2d_w}{c}\right)\left(\frac{2d_w}{c}\right)$$

So, the S_{11} should be convolved with this function to show the gated S_{11} (negative frequencies may be important, so remember that $S_{11}(-f) = S_{11}{}^*(f)$). To locate the proper time values, it will be required to see the time domain $S_{11}(t)$ as well. There are two expressions that are commonly used in network analyzers.

4.6.1 Low-Pass Transform

It may very often be desirable to truncate S_{11} after a certain *fmax*, and the inverse Fourier transform calculated using the fast Fourier transform (FFT) algorithm. Such truncation is usually not done abruptly, but a more gradual windowing is used—for example, $S_{11}(f)$ may be multiplied by $w(f)$, which is shown in Figure 4.19.

For FFT, $S_{11}(f)$ has to be discretized; say, $f = -N \Delta f$ to $N \Delta f$ are the discrete frequency values.

A typically used IFT is

$$S_{11}(t) = \frac{\Delta f}{W_0} \sum_{n=-N}^{N} S_{11}(n\Delta f)w(n\Delta f)e^{jn2\pi\Delta ft}$$

where

$$W_0 = \Delta f \sum_{n=-N}^{N} w(n\Delta f)$$

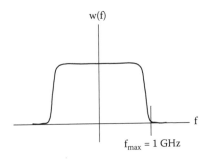

FIGURE 4.19
A sample window function.

There is a normalization built in here such that S_{11} = a constant 0 dB converts to a peak value of 1 in the impulse response (in the continuous frequency/time case it is an impulse).

4.6.2 Band-Pass Transform

Many systems have an inherently narrowband response; for example, a patch antenna may work only in the range 10 to 10.2 GHz. In this case, it will not be meaningful to compute the time domain response for 0 to 10.2 GHz. This is frequently implemented in a way that gives a somewhat unusual result. In real systems, $S_{11}(f)$ always displays the property $S_{11}(-f) = S_{11}^*(f)$. And consequently, if we take the IFT from any $-f_{max}$ to f_{max}, we will get a real $S_{11}(t)$. But in VNAs, frequently the IFT is taken only over positive frequencies f_{min} to f_{max}, yielding a complex $S_{11}(t)$ with a magnitude and phase. The phase, in particular, is hard to interpret. In any case, since usually the time domain response is only used to estimate the proper time interval for gating, this works satisfactorily.

The relevant IFT is now

$$S_{11}(t) = \frac{\Delta f}{W_0} \sum_{n=-N}^{N} S_{11}(f_c + n \ \Delta f) w(n\Delta f) e^{j(f_c + 2\pi n\Delta f)t}$$

$2N\Delta f$ is now the bandwidth.

The above description is, of course, very rudimentary. A more detailed description can be found in Dunsmore (2012) and VNA manuals.

4.7 Material Property Measurement Using the VNA

Another common use of the VNA is in the measurement of permittivity and permeability of materials. While different approaches are used for this (Baker-Jarvis et al., 1990; Blackham and Pollard, 1997), we will only give a brief description of the Nicolson-Ross-Weir (NRW) method. This was published in Weir (1974). A good overview of many of these techniques can be found in Agilent (2014a) or Chen et al. (2004).

The method works best for a sample loaded in a waveguide as shown in Figure 4.20, although a transverse electromagnetic (TEM) guide such a coaxial line can also be used.

The sample has to be in the shape that fits exactly inside the waveguide (for some materials such as ceramics or fabrics or films, this may be difficult). Now using a VNA, S_{11} and S_{21} are measured, ensuring that port 1 is at

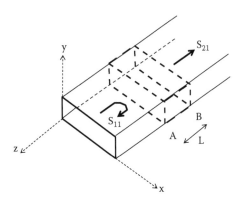

FIGURE 4.20
The NRW method of measuring ε_r and μ_r.

A and port 2 at B. Other than calibration, modern VNAs have the facilities of de-embedding or port extension (Agilent, 2014b), which can be used to ensure this.

These S-parameters are related to the material properties (complex ε_r and μ_r) through the following relations.

Define Γ:

$$\Gamma = \frac{\dfrac{\omega\mu_0\mu_r}{\beta} - \dfrac{\omega\mu_0}{\beta_0}}{\dfrac{\omega\mu_0\mu_r}{\beta} + \dfrac{\omega\mu_0}{\beta_0}} = \frac{\dfrac{\mu_r}{\beta} - \dfrac{1}{\beta_0}}{\dfrac{\mu_r}{\beta} + \dfrac{1}{\beta_0}} \quad \text{where} \quad \beta = \sqrt{\omega^2\mu\varepsilon - \left(\frac{\pi}{a}\right)^2} \qquad (4.50)$$

β is the complex propagation constant in the sample, and β_0 is the propagation constant in the waveguide outside the sample for which $\mu = \mu_0$ and $\varepsilon = \varepsilon_0$. Γ is actually the reflection coefficient at the air-dielectric interface (A in Figure 4.20) if the dielectric is infinite in length. Using transmission line theory, Γ can be related to the S-parameters:

$$\Gamma = X \pm \sqrt{X^2 - 1} \quad \text{where} = \frac{S_{11}^2 - S_{s1}^2 + 1}{2S_{11}}$$

The correct + or − sign is chosen to ensure that $|\Gamma|_r < 1$.

Having determined Γ, (4.50) can be solved for (μ_r/β) to give

$$\frac{\mu_r}{\beta} = \frac{1}{\beta_0}\frac{1+\Gamma}{1-\Gamma} \qquad (4.51)$$

Before this can be used to solve for μr, some definitions will be useful:

λ_0: Free-space wavelength $= c/f$

λ_{g0}: Wavelength in the air region of the waveguide $= 2\pi/\beta_0$

Coming back to (4.51) we see that before evaluating μ_r, β has to be known. For this we have the additional relation from P that is defined as $e^{-j\beta L}$. β here is complex.

Again, from transmission line theory,

$$P = \frac{S_{11} + S_{21} - \Gamma}{1 - (S_{11} + S_{21})\Gamma} \tag{4.52}$$

So, P can also be evaluated. From the definition of P:

$$\beta = \ln(P)/(-jL) \tag{4.53}$$

It appears that the solution is complete; unfortunately, a problem arises. Since P is complex, let its magnitude and phase be r and θ, i.e., $P = re^{j\theta} = re^{j\theta + 2n\pi}$. Then $\ln(P) = \ln(r) + j(\theta + 2n\pi)$. The choice of n is not clear. So, we will end up with many values: $\beta_1, \beta_2, \ldots$. For each of these we can calculate a value of μ_r from (4.51), giving us $\mu_{r1}, \mu_{r2}, \ldots$, and further, using the definition of β in (4.50) we can calculate $\varepsilon_{r1}, \varepsilon_{r2}, \ldots$. Which is the correct set of values?

To answer this we look at group delay, defined as $\tau_{meas} = \frac{-1}{2\pi}\frac{d}{df}(\angle S_{21})$. This is, of course, directly available from the VNA. It can also be calculated *approximately* (ignoring multiple reflections in the dielectric) as

$$\tau_{calc} = \frac{d\phi}{d\omega}, \text{ where } \phi = Re(\beta) \times L$$

$$\Rightarrow \tau_{calc} = L \times Re\frac{d\beta}{d\omega} = L \times Re\frac{\omega\mu\varepsilon}{\beta} \tag{4.54}$$

using the definition of β in (4.50) and assuming that material properties do not change sharply with frequency. Using (4.54), τ_{calc} can be evaluated for each set $(\beta_i, \mu_{ri}, \varepsilon_{ri})$, and the value that is closest to the measured group delay is identified as the correct one. This completes the determination of complex material properties.

The method will fail when the frequency is such that L is a multiple of half wavelength in the material. In this case, $S_{11} \cong 0$ (it is exactly 0 for lossless materials and most materials of interest are only slightly lossy) and problems arise with calculating X.

4.8 Nonlinear Measurements and X-Parameters

As we have seen, S-parameters are defined only for linear time-invariant circuits. It is tempting to extend the concept of relating incident and scattered waves to the nonlinear domain. Suppose that instead of just ai representing an incident wave at port i, we now have a double suffix a_{mn} representing the incident wave at the mth port, but at a frequency $= nf_0$, where f_0 is the fundamental frequency. Similarly, we may define b_{pq}. Can we have the following relation?

$$b_{pq} = \sum_{m=1}^{M} \sum_{n=-N}^{N} S_{pqmn} a_{mn}$$

The answer is no. Such a simple relation totally misses the nonlinearity. Nonlinear effects are unfortunately too complicated to allow for a general theory. It is only by making some drastic assumption that we can develop a mathematical framework.

The assumptions made in the polyharmonic distortion (PHD) (Verspecht and Root, 2006) technique are:

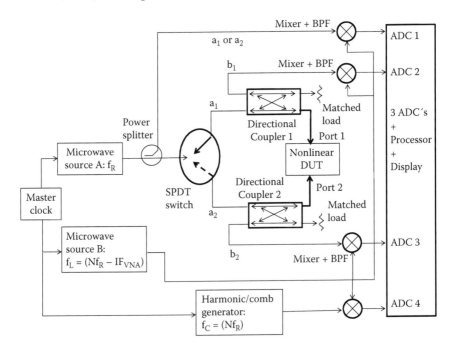

FIGURE 4.21
Simplified NVNA architecture.

1. A single f_0 and its harmonics exist—so transients are not considered (recently, the technique has been extended to more than one fundamental as well).

2. The fundamental at port 1 is the dominant signal—all other a_{mn} are much smaller.

Now, indeed, we can write a relation between b_{pq} and a_{mn}. This is

$$b_{pq} = X^F_{pq}(|a_{11}|)P^q + \sum_{m,n=1,0}^{M,N} X^{(S)}_{pqmn}(|a_{11}|)P^{q-n} a_{mn} + \sum_{m,n=1,0}^{M,N} X^{(T)}_{pqmn}(|a_{11}|)P^{q+n} a^*_{mn}$$

where $P = \exp(j \angle a_{11})$.

This contains the *nonlinear complex functions of one real variable*: XF, $X^{(S)}$, and $X^{(T)}$. This is of course a huge improvement from nonlinear functions of multiple variables (a_{mn}). The above assumptions have resulted in this simplification. Considering typical numbers, two 2 ports, and up to the fourth harmonic (so, five harmonics, including dc), there are 10 functions XF, 100 functions $X^{(S)}$, and 100 functions $X^{(T)}$. Fortunately, these functions are usually quite smooth and well behaved. Agilent Technologies has developed a technique for efficiently acquiring these functions from terminal measurements on a supplied component—this capability is found in the latest nonlinear VNA (NVNA) products from the company. Once these functions have been acquired, circuit simulation-based optimization may be carried out to optimize parameters such a power-added efficiency, harmonic distortion, etc.

The NVNA actually requires only one component extra, compared to the normal VNA—the rest is a matter of software. This extra component is a multifrequency phase reference that must replace source B in Figure 4.6 or Figure 4.14. Now that different harmonics are considered, we are faced with the issue of measuring, say, a_{11} and b_{22}, *including phase*. Using the normal VNA (say, $IF_{VNA} = 10$ MHz), assuming $f_0 = 1$ GHz, this can be done by first measuring a_{11} with source A at 1 GHz, and source B generating an LO at, say, 0.99 GHz, followed by source A at 2 GHz and source B generating LO at 1.99 GHz for b_{22}. Unfortunately, there is no known phase relationship between these two LO signals, so any measured phase difference between a_{11} and b_{22} will be meaningless. For this approach to be successful, we need to know the differences between the LO phases when it is operating at 0.99 GHz and when it is operating at 1.99 GHz. This can be done by providing a separate reference signal generating 1 GHz and its harmonics (say, a comb generator at 1 GHz). Now the LO at 0.99 GHz will mix with the fundamental of this to produce 0.01 GHz, and when the LO is switched to 1.99 GHz, it will mix with the second harmonic from the comb generator to again yield another 0.01 GHz signal. The phase difference between these two 0.01 GHz signals will yield the phase relationship between the two LO signals that we were looking for.

A conceptual block diagram for this NVNA is shown in Figure 4.20. More information can be found in Vye (2010).

It is seen from Figure 4.20 that only an external harmonic generator is required. Although we have discussed only two-port VNAs, many instruments are designed for use with three- or four-port components. For a four-port (linear) VNA the switch will become an SP4T (single-pole four-throw) switch, and two more sets of coupler + mixer + ADC will be required. To further convert such a four-port VNA into a three-port NVNA, the fourth mixer + ADC may be used with an external harmonic generator as shown.

This technique is still in its infancy—publications regularly appear on this topic in technical journals. Also, while we have focused on the X-parameter approach popularized by Agilent Technologies, by no means is this the only way to approach the topic of nonlinear component characterization. Other techniques have also been developed—a good overview of these can be found in Vye (2010).

4.9 Conclusion

An overview of the vector network analyzer is presented, explaining its principle of operation and calibration. Material property measurement is also discussed briefly. Extension of this instrument into the nonlinear domain is then introduced. There is, of course, abundant information on this topic available in the public domain literature—interested readers should definitely study some of them.

Problems

1. A two-port has S-parameters (50 Ω reference):

$$\begin{bmatrix} 0.5\angle 0° & 0.8\angle 0° \\ 0.8\angle 0° & 0.5\angle 0° \end{bmatrix}$$

S_{21} is measured using a 70 Ω source at port 1, and a downconverter + ADC at port 2, which has input impedance 50 Ω. There is a separate source power splitter + downconverter + ADC that correctly supplies the available source power and the source phase for reference.

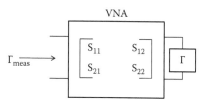

VNA

$$\begin{bmatrix} S_{11} & S_{12} \\ S_{21} & S_{22} \end{bmatrix}$$

$\Gamma_{meas} \longrightarrow$

Γ

FIGURE 4.22
One-port calibration using different notation.

Two measurements are done, one with the network and one without it (direct connection from source to downconverter), and the ratio taken. What S_{21} will be measured?

2. It is required to do one-port calibration of a VNA to measure the reflection coefficient of a load. If the true value is Γ, the measured value is $\Gamma meas$. The model used is

Here the shown S-parameters are the unknown coefficients that have to be determined through calibration. Modify the SOLT (one-port) calibration—now instead of e_{00}, etc., we have $[S]$.

3. A directional coupler, microwave source, matched terminations, and a short circuit are used to measure S_{11} of an unknown load. Downconverters and ADCs are available to make vector measurements as in Figure 4.4. The measured S_{11} turned out to be $0.3 - j0.2$. What is the true S_{11} if the reflection coefficient of the short is actually $-0.9 - j0.1$ and there are no other imperfections?

4. A one-port 50 Ω VNA calibration is being done for the following situation:

The forward and reflected waves (downconverted) that are sampled and measured are denoted by a_0 and b_0 as usual.

The forward and reflected RF signals at port 1 are denoted by a_1 and b_1.

The usual coupler mismatch and nonideal isolation are present.

There is an unknown length of 50 Ω lossless transmission line from port 1 to the actual load—the effect of this line is to be removed.

Draw the signal-flow graph for this situation, and suggest how the usual one-port calibration procedure can be modified (if any modification is required).

5. Suppose that the LO frequency of a NVNA is $\cos(2\pi (0.99)10^9 t + \theta)$ for the fundamental and $\cos(2\pi (1.99)10^9 t + \phi)$ for the second harmonic using a VNA IF of 10 MHz. Write the reference harmonic waveforms and ensure that the difference $\theta - \phi$ is indeed evaluated.

References

Agilent. *Applying Error Correction to Network Analyzer Measurements.* Application Note 1287–3. 2002.

Agilent. *Basics of Measuring the Dielectric Properties of Materials.* Application Note, 2013a.

Agilent. *Agilent FieldFox RF Vector Network Analyzer N9923A User's Guide,* 2013b.

Baker-Jarvis, J., et al. Improved Technique for Determining Complex Permittivity with the Transmission/Reflection Method. *IEEE Transactions on Microwave Theory and Techniques,* 38(8), 1990.

Blackham, D.V., and R.D. Pollard. An Improved Technique for Permittivity Measurements Using a Coaxial Probe. *IEEE Transactions on Instrumentation and Measurement,* 46(5), 1093–1099, 1997.

Chen, L., C.P. Neo, C.K. Ong, V.V. Varadan, and V.K. Varadan. *Microwave Electronics: Measurement and Materials Characterization.* John Wiley & Sons, Chichester, UK, 2004.

Dunsmore, J.P. *Handbook of Microwave Component Measurements.* John Wiley & Sons, Chichester, U.K, 2012.

Engen, G. *Microwave Circuit Theory: And Foundations of Microwave Metrology,* Peter Peregrinas, Ltd., London, U.K., 1992.

Engen, G.F. and C.A. Hoer. Through-Reflect-Line: An Improved Technique for Calibrating the Dual Six Port Automatic Network Analyzer. *IEEE Transactions on Microwave Theory and Techniques,* December 1979.

Ha, T.T. *Solid State Microwave Amplifier Design.* John Wiley & Sons, New York, 1981.

Verspecht, J., and D.E. Root. Polyharmonic Distortion Modeling. *IEEE Microwave Magazine,* June 2006.

Vye, D. Fundamentally Changing Nonlinear Microwave Design. *Microw. J.* (cover feature), March 2010.

Weir, Automatic Measurement of Complex Dielectric Constant and Permeability at Microwave Frequencies. *Proceedings of the IEEE,* 62(1), 1974.

5

Spectrum Analyzer

A spectrum analyzer is one instrument for gathering information about a signal. Unlike the network analyzer (where the signal frequency is set by the user), information is not available about this signal before measurement. It is obvious that the time domain waveform contains complete information about a signal. But it is not always that all time domain measurements are useful—as will be seen later, time domain measurement using an oscilloscope has its own limitation. Measurement of spectra (power spectral density, for example, which contains only partial information about a signal) also has an equal importance. One direct advantage is measuring the electromagnetic emission. With the overcrowding of the spectrum today, including modern numerous wireless standards like 2G, 3G, EDGE, GSM, LTE, WiMax, etc., in the same frequency band, it is imperative that emissions from the carrier under test do not significantly interfere with signals falling close to the same. One particular example is the emission from your own cell phone interfering with other handsets reusing the same frequency band. The result is something that you would not want, quantified as poor reception and cross talk. Another form of distortion that manifests itself is the distortion of the message modulated onto the carrier. Third-order intermodulation (two tones of a complex signal modulating each other) can be particularly troublesome because the distortion components can fall within the band of interest. This is particularly dangerous as using low-pass filters is not appropriate; the desired signal and the unwanted signal share the same frequency space.

Another important application is the monitoring of spectrum. Various radio services, including emergency and cellular services, are allocated parts of the frequency band by the government, and it is understood that the interaction between them is kept to a minimum. However, in reality it may not always be possible. There are multiple cases where two emergency services have to share frequency bands very close to each other. In such cases, determination of the spectra from both of these channels is important. This helps the monitoring agency to figure out how much signal from, say, channel A is leaked or spilled over into channel B, and vice versa, when both are simultaneously transmitting. Needless to say, the greater the spill, the more is the interference and the greater are the chances of a poorer transmission and reception. In technical terms, this parameter is termed the adjacent channel power leakage ratio (ACPR) and is slowly becoming the gold standard for characterization of transmitters under modulated signals following the most modern standards.

As another example of **electromagnetic interference** (EMI), the motherboard of your laptop is a source of EMI. All manufacturers are therefore bound to test their equipment after manufacture for emission levels vs. frequency and make sure that these emissions are within acceptable government standards.

Some examples with real spectrum analyzer data follow (Figures 5.1 to 5.4) to determine the usefulness of such an instrument in the characterization of the earlier discussed points. These are taken from Application Note 150 (*Spectrum Analysis Basics*) from Agilent Technologies. Most of the material in this chapter is adapted from this very detailed application note.

5.1 Common Measurements Using the Spectrum Analyzer

Common spectrum analyzer measurements include frequency, power, modulation, distortion, and noise. The information contained in the signal is important especially for systems that have a limited bandwidth. It is also important to characterize the transmitted power. Low power essentially means that the signal may not reach its destination. Again, transmitting more power than required may drain the batteries, create distortion, and cause unwanted thermal effects. Another important quantity is measuring the quality of modulation in the transmitted signal, quantifying how much information is lost in transmission. Analog modulation measurements include tests on modulation degree sideband amplitude, modulation quality, and occupied bandwidth. Digital modulation measurements include IQ (in phase/quadrature) imbalance, error vector magnitude (EVM), and phase error vs. time.

Measuring distortion at both the transmitter and the receiver is critical in communication systems. Excessive harmonic distortion at the output of a transmitter would cause unwanted interference with other bands. For a receiver again, the intermodulation distortion is important, as it causes signal cross talk. A very common example is the intermodulation in cable TV receivers where a distortion in one particular channel may cause significant distortion in all other channels. Spurious emissions, harmonic, and intermodulation distortion measurements are commonly carried out using the spectrum analyzer. Another very important component of measurement is noise. Noise is ubiquitous; any lossy or active electronic component generates noise. The spectrum analyzer is useful to characterize this excess noise in terms of noise figure, which is defined as the ratio between the signal-to-noise ratio at the input and the signal-to-noise ratio at the output. The overall noise performance of a system has become a

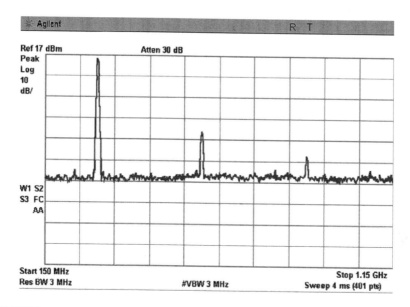

FIGURE 5.1

Harmonic distortion characterization. (c) Copyright Agilent Technologies 2004-5. Reproduced with permission.

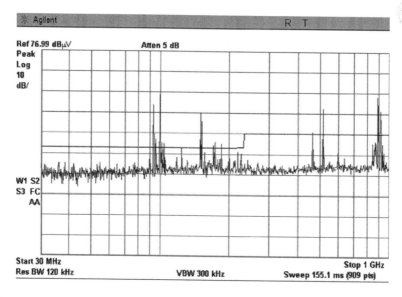

FIGURE 5.2

EMI measurements. (c) Copyright Agilent Technologies 2004-5. Reproduced with permission.

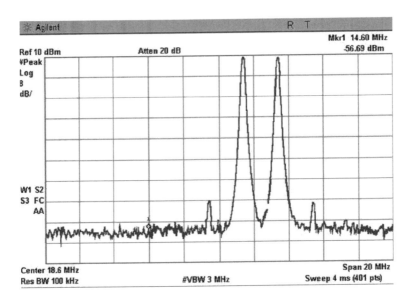

FIGURE 5.3
Intermodulation test of an RF power amplifier. (c) Copyright Agilent Technologies 2004-5.
Reproduced with permission.

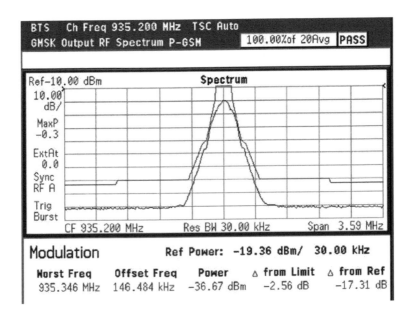

FIGURE 5.4
GSM radio signal and spectral mask showing limits of unwanted emissions. (c) Copyright
Agilent Technologies 2004-5. Reproduced with permission.

critical parameter in characterization of a complete transceiver module. More details on spectrum analyzer-based measurements are given at the end of this chapter.

5.2 Types of Signal Analyzers

The swept tuned super-heterodyne spectrum analyzer is the one that you would commonly see in the lab. An important non-super-heterodyne analyzer is the Fourier analyzer. This digitizes the input time domain signal and then, using standard digital signal processor (DSP) techniques, converts it into the frequency domain after performing a fast Fourier transform. This helps to measure both phase and magnitude of the incoming signal. However, these analyzers are only available at baseband until 40 MHz and have a poor selectivity and dynamic range compared to the conventional super-heterodyne analyzer. The Fourier analyzer's counterpart in the radio frequency (RF) domain is the vector signal analyzer (VSA), which extends the frequency range by using downconverters in front of the digitizer. They offer fast, high-resolution spectrum measurements, demodulation, and advanced time domain analysis, including ACPR measurements and EVM. They are especially useful for characterizing complex signals such as burst, transient, or modulated signals used in communications, video, broadcast, sonar, and ultrasound imaging applications.

Two kinds of analysis have become clear from our discussion: one is a spectral analysis and the other is a vector signal analysis. However, with the advent of digital technology and digital signal processing techniques, this distinction is slowly becoming blurred. Earlier, the digitization was done at the video frequencies at tens of KHz. Video signals did not contain any phase information; hence, only magnitude information was displayed. With the new complex transmitted signals, this digitization is now done at the instrument's input or after amplification or after the first downconverter. This preserves both phase and magnitude information. Additional DSP functions are then implemented from either the firmware of the analyzer or a computer connected to the analyzer, which helps in carrying out true vector measurements. Figure 5.5 shows an example vector measurement using the constellation diagram of a real-time transmitted QPSK (quadrature phase shift keying) signal.

In this chapter we will discuss only traditional spectrum analysis; more modern vector signal analyzers are described in detail in Agilent Technologies Application Note 150-15.

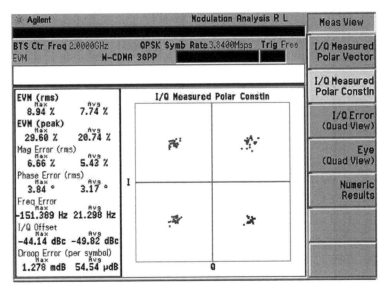

FIGURE 5.5

Constellation and EVM measurements using a modern vector signal analyzer. (c) Copyright
Agilent Technologies 2004-5. Reproduced with permission.

5.3 Basic Idea Behind Spectrum Analyzers

The spectrum analyzer measures power spectral density, as described in
Chapter 2. Following the concept explored in Section 2.2, the operation of a
spectrum analyzer can be described as shown in Figure 5.6. Note that only
single-sided spectra (so power values are double the true double-sided spectra) are considered in this chapter.

As discussed in Section 2.2, if we plot the power dissipated in the load R,
as a function of the center frequency f (i.e., we plot $P_L(f)$), what we get is the
power spectral density of the power delivered to the load R, multiplied by
the narrow bandwidth B. Strictly speaking, this is one-fourth of the power
spectral density of $V(t)$ feeding a load R, the way we defined it in Section 2.2
(here $V(t)$ sees two resistances R: e.g., the 50 Ω resistance of the source and
the 50 Ω spectrum analyzer input impedance). In microwave metrology, the
present definition is what the display of a spectrum analyzer follows. The
student may work out the relation between the displayed power spectral
density and the available power density if the source impedance is some
general Z instead of R (same as spectrum analyzer input resistance).

An example will clarify the operation of the instrument shown in
Figure 5.6. Let us say that the input $v(t)$ is a single-tone signal corrupted by
amplitude modulation (AM) noise, as discussed in Section 2.2. Its power
spectral density (available power) is shown in Figure 5.7. The band-pass filter
$|S_{21}|$ is shown in Figure 5.8.

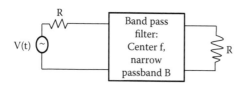

FIGURE 5.6
Conceptual spectrum analyzer.

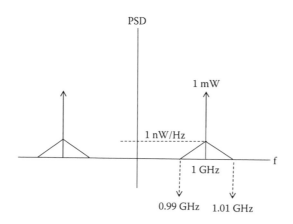

FIGURE 5.7
An example input to a spectrum analyzer.

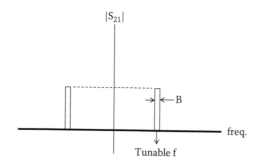

FIGURE 5.8
Transfer function of the band-pass filter.

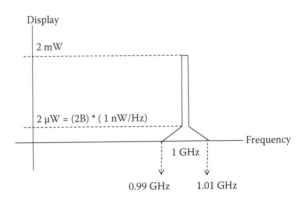

FIGURE 5.9
Spectrum analyzer display.

Following our earlier convention, all spectra are shown as double-sided, although the final display is always single-sided.

In this case, if we sweep f from 0 to high frequencies in Figure 5.7, and assuming that $B = 1$ KHz, say, is small compared to the 0.02 GHz noise bandwidth, we get the display shown in Figure 5.9.

Notice that the 2 µW level is proportional to the filter bandwidth (called resolution bandwidth (RBW)), while the 2mW level is not, because of the impulse function (concentration of finite power at a spot frequency). Strictly speaking, the display will have smoothed corners at 0.99 and 1.01 GHz, and also at 1 GHz $\pm$ $B/2$. The top of the curve will also show a small rooftop shape—these details are left to the student.

An actual filter shape will not be so abrupt—a Gaussian shape is more realistic. This leads to the familiar spectrum analyzer displays like Figure 5.1.

Even with such a realistic shape, a filter with a constant bandwidth and a wide-ranging swept center frequency is not practical, and an (almost) equivalent operation has to be achieved in a more complicated way.

A very important feature of the spectrum analyzer is its noise immunity compared to an oscilloscope; this is because the narrow resolution bandwidth allows only a minuscule noise power. Notice that it is *not* a real-time instrument—for measuring the spectrum of, say, a single pulse (not periodic), a traditional spectrum analyzer will not work.

5.4 Building Blocks of a Spectrum Analyzer

Figure 5.10 is a simplified representation of the internal structure of the spectrum analyzer based on the super-heterodyne architecture. The term *super* refers to frequencies far greater than audio frequencies, and *heterodyne*

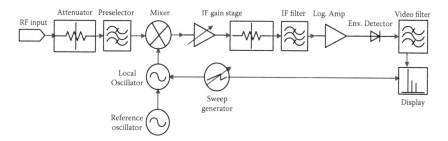

FIGURE 5.10
Basic block diagram of a spectrum analyzer.

refers to mixing. An input signal passes first through an attenuator, and then is filtered to remove potential spurious signals. This then serves as an input to a mixer where the other input comes from a local oscillator (LO). This is a synthesized signal generated from a highly stable crystal oscillator reference (older equipment used an analog swept frequency source or voltage-controlled oscillator [VCO]—more information on this is given in Chapter 7). The mixer then produces the sum and difference frequencies along with their harmonics and possibly the original signals and their harmonics. All these frequency components are then further processed or rejected through the intermediate frequency (IF) filter, rectified by the envelope detector, digitized, and displayed. The display in older equipment is essentially a cathode ray tube, and the horizontal sweep is carried out by a ramp generator, which also generates the different LO frequencies. The output may be understood as follows. The display grid is essentially a horizontal and a vertical with 10 divisions each. The horizontal axis is an indicator of frequency with an increase from left to right. There are two ways to set the frequencies. We can either set the center frequency followed by the span on the analyzer or set the start and the stop frequency. The vertical axis is an indicator of amplitude calibrated on either a linear or a logarithmic scale in dB. The log scale display automatically leads to a higher dynamic range. There is also the provision of a delta marker, which enables the user to read off values on both axes at multiple frequency points at a time. There is a control present to calibrate both the frequency and the amplitude scale.

Let us now analyze the individual blocks shown in Figure 5.10.

5.4.1 RF Attenuator

The mixer, being a nonlinear component, is susceptible to compression if the presented input signal is high. The attenuator helps set precisely the right level of the input signal to prevent overload, gain compression, and distortion to the signal passing through the mixer. The attenuation may be set automatically depending on the reference signal level or may be manually tuned in steps of 10, 5, 2, and 1 dB. There is a blocking capacitor placed before the attenuator that ensures that the analyzer is not damaged by dc. However, this also leads to attenuation of very low frequency signals, and hence typically the frequency range of all analyzers begins from tens of KHz and not dc.

5.4.2 Low-Pass Filter or Preselector

The primary function of the low-pass filter (LPF) is to block high-frequency signals, and hence out-of-band signals, from reaching the mixer and creating unwanted products at the output. However, in modern spectrum analyzers this is replaced by a tunable filter, also known as a preselector, which tunes its pass-band to exactly the same range in which one wishes to view the signal. This will be described in detail later.

5.4.3 Tuning the Analyzer

The tuning is primarily effected by the center frequency of the IF filter, the frequency range provided by the LO, and the frequencies reaching the mixer after being low-pass filtered. The most prominent components coming out of the mixer are the sum and difference of the input and LO frequencies. Now let us say that we can work up an arrangement such that the desired signal falls above or below the LO by the IF; then we understand that one of the mixing products would definitely fall within the pass-band of the IF filter, and hence be detected to create an amplitude response on the display.

For example, let us say that our desired frequency range is from 0 to 3 GHz.

1. Choose a suitable IF preferably within this band, say, 1 GHz.
2. The output of the mixer also includes the original signal; hence, a 1 GHz LO signal will automatically pass through the entire filter chain and be displayed at the output, regardless of any input.
3. This independence from the input is not advisable, as it leads to a gap in the displayed frequency spectrum: the amplitude of this signal is independent of the input.
4. Thus, the choice quoted in step 1 is not useful.
5. To eliminate the issue with a low IF, let us now choose an IF that is outside the range of the spectrum analyzer. In this example let us start with 3.9 GHz.
6. If we now start tuning the LO from the IF, and end at 3 GHz above the IF, we would be able to cover the entire frequency band. This leads to a resultant equation for the LO frequency:

$$f_{LO} = f_{sig} + f_{IF} \tag{5.1}$$

If we now wish to tune our analyzer to 1 KHz, 1.5 GHz, or 3 GHz, we simply need to increase our LO frequency according to (5.1), the highest LO frequency being 6.9 GHz. Of course, the displayed output when the instrument is tuned to receive any of these (1 KHz, 1.5 GHz, or 3 GHz) depends on the strength of the input *at that frequency*—in many cases a nonzero value will be displayed at exactly one frequency (so the input is single tone).

7. Now let us consider a case where the input signal is at 8.2 GHz or outside the range of the analyzer. Now when the LO sweeps from 3.9 to 6.9 GHz, at 4.3 GHz it is exactly an IF away from 8.2 GHz. This leads to a mixing product that exactly falls within the IF band, and hence is forwarded to the display. This also means that (5.1) could have been expressed as

$$f_{LO} = f_{sig} - f_{IF} \tag{5.2}$$

This also means another tuning range for the same analyzer, from 7.8 to 10.9 GHz.

8. However, it is the job of the LPF discussed earlier to prevent these higher-frequency signals from reaching the mixer and consequently the display. These signals also include the IF. Hence, it is desirable that the LPF's cutoff frequency is kept below the IF.

Sometimes to resolve closely spaced signals, it is necessary to set the BW of the IF filter as low as 1 Hz. However, it may not be possible to set such kinds of numbers at 3.9 GHz. Hence, a multilevel downconversion is employed. Figure 5.11 illustrates the downconversion. The complete tuning equation for such a multistep implementation is

$$f_{sig} = f_{LO1} - (f_{LO2} + f_{LO3} + f_{FINALIF}) \tag{5.3}$$

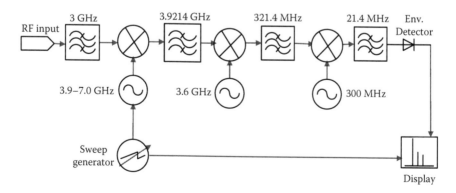

FIGURE 5.11
Multiple downconversion stages in a spectrum analyzer.

However,

$$f_{LO2} + f_{LO3} + f_{FINALIF} = 3.9214 \text{ GHz}$$

Surprisingly, this is the first IF. Hence, (5.3) may be simplified using only the first IF (3.9214 GHz). So, the relevant equation is $f_{LO} = f_{sig} + 3.9214 \text{ GHz}$, and the LO sweeps from 3.9214 to 6.9214 GHz, tuning an RF input range 0 to 3 GHz, frequency point by point.

At the beginning of each sweep, some LO signal leaks and appears at the output of the mixer at 3.9214 GHz. This translates from (5.1) as an input signal at 0 Hz. This is commonly termed dc feed-through in a spectrum analyzer, and the LO being strong, amplitude of this signal is obviously quite large. Thus, signals very close to dc would be masked; hence, the display range in a typical analyzer starts from a few kHz and not dc (even if there was no dc block at the input).

5.4.4 IF Gain Block

This is typically a variable gain amplifier. It is primarily used to adjust the vertical scale without actually affecting the amplitude of the displayed signal. When the IF gain is changed, the reference level is auto-adjusted to maintain the correct displayed level of the signal. With change in the input RF attenuator, the IF gain is auto-adjusted.

This architecture would appear to give a rather limited upper frequency limit. Later we will see how to extend the upper frequency limit of the spectrum analyzer.

5.5 Features of the Spectrum Analyzer

Frequency resolution is defined as the ability of an analyzer to separate two closely spaced signals into two distinct responses. In a traditional spectrum analyzer the input frequency (or frequencies) is fixed and the LO frequency is swept. The mixer generates products that are filtered by the IF filter. The mixing products are hence obviously swept with frequency. When the mixing product sweeps past the designated fixed IF, the shape of the band-pass filter is sketched on the display as discussed in the example earlier. The narrowest filter in the chain determines the overall bandwidth of the displayed signal. Hence, two signals should be suitably spaced in frequency to prevent the overlap of responses. However, modern spectrum analyzers have selectable IF filters (often through DSP), which leads to very narrow IF BW, and hence an effective resolution of two closely spaced signals. A useful number to characterize this spacing between these signals is the 3 dB BW of the IF filter. For equal amplitude signals there would be a 3 dB dip between the two peaks traced out by the signal. For nonequal amplitude cases, two signals very close to each other in frequency may be difficult to resolve if the smaller-amplitude signal gets lost in the skirt of the larger signal. Bandwidth selectivity is another specification of a resolution filter for unequal sinusoids. This is defined as the ratio of the 60 dB to the 3 dB BW, and typical IF filters assume a Gaussian shape.

5.5.1 Residual FM

Another additional factor that determines the resolution of the signals is the residual frequency modulation (FM) signal from the instability of the LO signals. Earlier the residual FM of the yttrium-iron-garnet (YIG) LO signals could go as high as 1 KHz. This signal was transferred to the mixer, and consequently the mixing products were unstable. The BW of the IF filter (the RBW) was limited (on the low side) by the maximum residual FM signal. However, modern analyzers have now reduced this to a few Hz, and hence have consequently improved the RBW. Any instability on the display is now purely attributed to the quality of the input signal and not the LO signal.

5.5.2 Phase Noise of the LO Signal

The LO frequency or phase instability is sometimes directly transferred to the display, as no oscillator signal is ideal. It is both amplitude and frequency modulated by random noise. These noise sidebands appear at the output above the noise floor and go further and further down if the LO becomes more and more stable. Another way of avoiding these noise bumps is by reducing the RBW of the IF filter. If we reduce the resolution bandwidth by a factor of 10, the level of the displayed phase noise decreases by 10 dB. The phase-locked loops applied to stabilize the LO play a major role in deciding the shape of the displayed phase noise. The phase noise of the analyzer is visible only under very low RBW settings when the noise obscures the skirt of the filter. For higher RBWs, these noise bumps are hidden under the skirt of the IF filter. Phase noise is the actual bottleneck in resolving two signals of unequal amplitude in a spectrum analyzer as the lower-amplitude signal gets obscured under the skirt of the noise bumps.

5.5.3 Sweep Time

A direct trade-off with selecting a very low resolution IF filter is the sweep time. Sweep time directly affects the time it takes to complete a measurement. The IF filters are band-limited circuits with finite time to charge and discharge. If the mixing products are swept through them very quickly, there is a high possibility of a loss of information. The amount of time the mixing

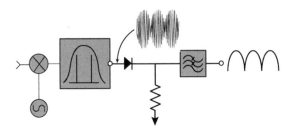

FIGURE 5.12
The output of the envelope detector follows the peaks of the signal. (c) Copyright Agilent
Technologies 2004-5. Reproduced with permission.

product spends within the pass-band of the filter is directly proportional to
the sweep time. A change in resolution hence has a dramatic effect on sweep
time. Roughly, the time spent at each frequency point in the LO sweep =
1/RBW. As typical numbers, with an RBW of 100 Hz, and 1000 frequency
points, each sweep will take 10 s.

5.5.4 Envelope Detector

The envelope detector helps to convert the IF signal to a video signal. It consists of a diode, a resistive load, and a low-pass filter (Figure 5.12). This is
very similar to the diode-based power sensor described in Chapter 3, but
usually operates at higher power levels. When the AM wave from the IF
filter is applied to this detector, it follows the changes in its envelope and
disregards the instantaneous value of the IF signal. During measurements
when we choose a very narrow RBW and fix the LO frequency to tune the
output to one of the spectral components of the signal, the output becomes a
sine wave with no variations in its envelope. The envelope detector then produces a dc signal as its output. In another case we may use a slightly higher
RBW, and two signals lie within a spectral region that is narrower than our
RBW. In such cases the envelope varies owing to phase variations between
the two sine waves. The BW of the RBW filter determines the maximum rate
at which the IF signal can change. This bandwidth also determines how far
apart two input sinusoids can be so that after the mixing process they will
both be within the filter at the same time. If we have a 21.4 MHz final IF and
a 100 KHz BW, two input signals separated by 100 KHz would produce mixing products at 21.35 and 21.45 MHz and would hence fall within the mixer
BW. The detector will now track the changes in the envelope created by these
two signals instead of a 21.4 MHz signal. If the input consists only of isolated
tones (typically a fundamental and harmonics), then the envelope will be
constant and there is no such complication, but modern communication signals will yield a range of frequencies even when the RBW is small, in which
case there will be envelope variation.

This situation, when we wish to make use of the envelope variation (and not be satisfied with the average power coming out from the IF filter) gives rise to various types of detection procedures (sample, peak, normal, etc.). This is an advanced topic and some details can be found in Agilent (2004).

5.5.5 Displays

Early on, all spectrum analyzer displays were purely analog. However, this posed serious problems with longer sweep times for narrow RBWs. A variable persistence CRT (cathode ray tube) was then introduced by Agilent to adjust the fade rate of the display. This was then updated with digital circuitry. The detector types were then updated to be compatible with the digital display. Since spectrum analyzers display the signal as well as its internal noise, it is imperative that the display be smoothened. An LPF is included prior to the digitization stage to achieve this (just after the envelope detector). The cutoff frequency of this is reduced to a point below the IF filter BW. The displayed signal thus is not able to follow the rapid envelope variations of the signal, and hence effectively displays an averaged version as in Figure 5.13. This particular method is useful while measuring noise, as the peak-to-peak variations of this noise are eliminated. The sweep time is also an inverse function of the video bandwidth when VBW (video bandwith) is set lower than RBW. This filter determines the video bandwidth.

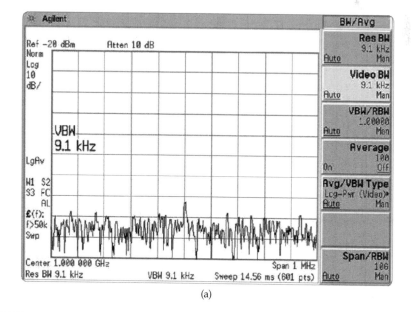

(a)

FIGURE 5.13

A reduced VBW smoothes the output signal considerably. (a) High video bandwidth. (b) Low video bandwidth. (*continued*)

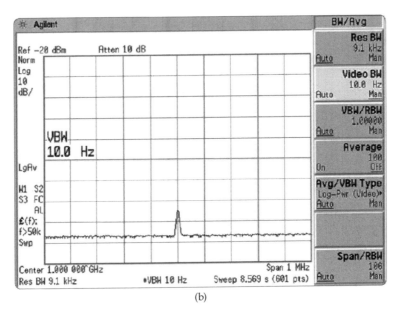

(b)

FIGURE 5.13

(*continued*) A reduced VBW smoothes the output signal considerably. (a) High video bandwidth. (b) Low video bandwidth. (c) Copyright Agilent Technologies 2004-5. Reproduced with permission.

5.5.6 Trace Averaging

Here averaging is established over two or more sweeps on a point-by-point basis. The averaging is specified over the number of sweeps as determined by the user. If we want to discern a low-level sinusoidal signal very close to noise, we may employ either video or trace averaging. However, there are differences between the two. Video averaging is carried out in real time, and each point is averaged only once, for a time of about 1/VBW on each sweep. Trace averaging, on the other hand, requires multiple sweeps to come up with the value of a particular displayed point. Hence, both of these methods will sometimes lead to distinctly different results.

5.6 Extending the Frequency Range

In the previous section we have seen when a spectrum analyzer is tuned to 3 GHz, all higher-order input terms are rejected by an LPF. A higher-input frequency range is also possible with the same analyzer. The easiest way to extend the range of this analyzer is by employing a concept of harmonic mixing. Recall that we had chosen our IF as 3.9 GHz, which is above the input range, as an IF within the input range was creating a hole in the displayed

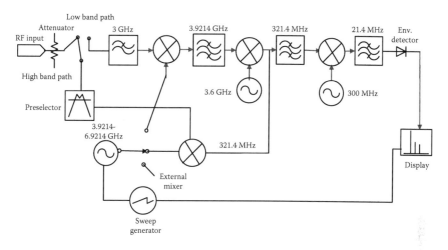

FIGURE 5.14
Switching arrangement for low and high bands.

frequency spectrum. In a typical case, let us choose the first IF for the higher-frequency range as 321.4 MHz. Notice that in the multiple downconversion stage for the original analyzer, the second IF is also at 321.4 MHz. Hence, all that needs to be done is to bypass the first IF mixer. Figure 5.14 shows a block diagram representation (the external mixer is explained later). Let us adopt a graphical approach to understand this multiband tuning in detail.

5.6.1 Low Band

Suppose we examine the basic equation of the mixer:

$$\pm(f_{sig} - f_{LO}) = f_{IF} \tag{5.4}$$

Or, rewriting this,

$$f_{sig} = f_{LO} \pm f_{IF} \tag{5.5}$$

This can be interpreted as f_{sig} is one of two linear functions of f_{LO}. If we plot the signal frequency along the Y-axis and the LO frequency along the X-axis, two straight lines will result (Figure 5.15). Any particular point on either of these lines, say, (x, y), signifies that when the LO reaches x during the sweep, an output indicating the power in input signal frequency y is displayed.

A mixing product at IF is obtained if the input signal is either above or below the LO signal by the IF signal. Therefore, we can determine the input frequency simply by adding or subtracting the IF from the LO frequency. To

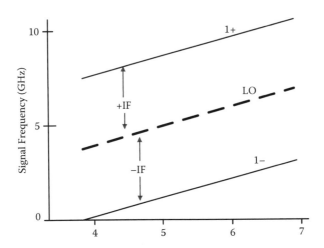

FIGURE 5.15
Tuning curves for fundamental mixing in the low-band, high-IF case.

determine the tuning range, we first plot the signal frequency against the LO frequency (the dashed line). When we subtract the IF from the dashed line, we get the desired tuning range of the LO. The label 1– indicates a first-order mixing term and the use of a minus sign in our original tuning equation (5.5). This graph can now be used to determine what LO frequency is required to receive a particular signal, or vice versa. The IF is 3.9214 GHz. For example, to display a 1 GHz signal, the LO must be tuned to 4.9214 GHz. Conversely, for an LO frequency of 6 GHz, the spectrum analyzer is tuned to a frequency of 2.0786 GHz. For the RF range 0 to 3G Hz, the LO range is 3.9214 to 6.9214 GHz. As discussed earlier, at the precise value of LO = 3.9214 GHz, corresponding to RF = 0, LO leakage through the mixer will lead to a spurious signal, which is often mistaken for a low-frequency spurious. So, care has to be taken when we expect a mixture of, say, GHz-level and KHz-level signals in the input.

The line above the dashed line in the graph (1+) is obtained for the higher band, namely, 7.8428 to 10.8428 GHz. These lines are separated by two times the IF. This upper band is automatically eliminated if we have a low-pass filter at the input.

5.6.2 Low-Band Path, Harmonic Mixing

It is well known that the LO provides a high-level input drive signal to the mixer. The mixer being a nonlinear device and under excitation by a large amplitude signal generates multiple harmonics of the LO frequencies. The

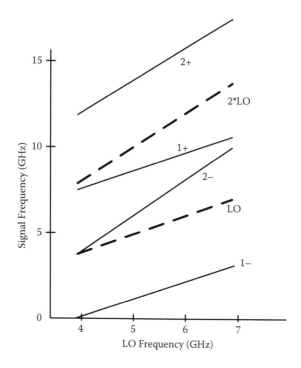

FIGURE 5.16
Signals in the 1– frequency range produce single, unambiguous responses in the low-band, high-IF case.

input signal mixes with these harmonics, and any mixing product that equals the IF is displayed. The tuning equation now becomes

$$f_{sig} = nf_{LO} \pm f_{IF} \tag{5.6}$$

Here n is the LO harmonic. To determine the tuning range, we again first plot the 2*LO frequency against the signal frequency (the upper dashed line) in Figure 5.16. The original LO signal also is displayed on the same graph. The lines 2+ and 2– refer to the IF added or subtracted, respectively, to the 2*LO curve. It is seen that these additional lines only come into action for input frequencies above 3.9214 GHz, while we have a low-pass filter that removes all signals above 3 GHz. So, harmonic mixing does not create problems.

5.6.3 High Band with Low IF (No Harmonic Mixing)

To determine the tuning range, we again first plot the LO frequency against the signal frequency (the dashed line) in Figure 5.17. Now the

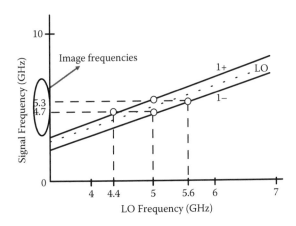

FIGURE 5.17
Tuning curves for fundamental mixing in the high-band, low-IF case.

1+ and 1– lines are much closer to each other; in fact, there is clear ambiguity (same RF sensed with different LO) due to the low IF of 321.4 MHz chosen. This close spacing of the tuning ranges complicates the measurement. For simplicity, let us round off the figures; now if we look at an LO frequency of 5 GHz on the X-axis it leads to two frequencies (4.7 and 5.3 GHz) on the signal Y-axis. Alternatively, if we look at 5.3 GHz on the Y-axis, it leads to values of 5 and 5.6 GHz on the X-axis, but also two simultaneous responses on the 1+ and 1– curves. This 5.6 GHz is twice the IF above 5 GHz. Now if we read the signal frequency graph at 4.7 GHz on the Y-axis, it leads to a 1+ response at an LO frequency of 4.4 GHz and a 1– response at an LO frequency of 5 GHz. The 4.4 GHz signal is again twice the IF below 5 GHz. Thus, it is possible for signals at different frequencies to produce the same output point on the display. As seen in the earlier example, input tones at both 4.7 and 5.3 GHz produce the same IF on the display when mixed with an LO signal of 5 GHz. These signals, termed image frequencies, are separated by 2*IF.

5.6.4 High Band with Low IF (with Harmonic Mixing)

The spectrum analyzer is now designed to operate among multiple frequency bands, using harmonics of the LO. The tuning curves are now shown in Figure 5.18. The frequency scale is calibrated for a specific LO harmonic depending on the band that we would tune to. Let us say in the 6.2 to the 13.2 GHz tuning range, the desired signal lies along the 2– curve. We now have an 11 GHz signal at the input. When the LO sweeps, multiple IFs are produced with the help of the 3+, 3–, 2+, and 2– curves. The desired response would fall on the 2– curve when the LO is tuned to 5.65 GHz: 11 GHz $= 2f_{LO} - 0.3$ GHz.

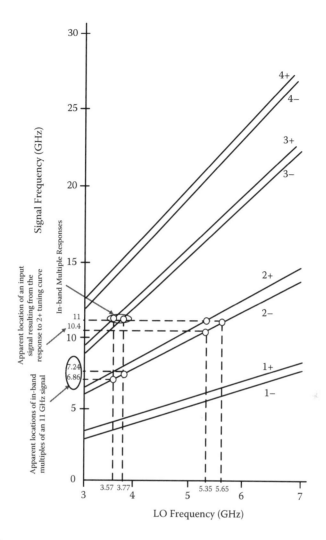

FIGURE 5.18
Tuning curves up to the fourth harmonic of LO showing in-band multiple responses to an 11 GHz input signal.

Looking at Figure 5.18, we can see that for this 11 GHz input, output will be detected when the LO is 3.57 or 3.77 GHz (if the LO sweep starts from below 3.9214 GHz for this high band) looking at 3+ and 3– curves. Also, we will have responses when LO reaches 5.35 and 5.65 GHz (2+ and 2– curves). Also, for LO values of 3.57, 3.77, and 5.35 GHz, it is possible that what we see on the display came from input frequencies of 6.86, 7.24, and 10.4 GHz, respectively. It is clear that any sort of manual identification of the correct signal will be difficult.

5.6.4.1 Preselector

Let us go back to our earlier example of having two signals at 4.7 and 5.3 GHz. One can always set up a band-pass filter to allow the desired signal into the analyzer and reject the other one. However, having a fixed filter does not always solve the problem of a crowded spectrum, and hence can lead to potential confusion. Thus, we use a tunable filter that adjusts its center frequency in such a way that tracks the frequency of the appropriate mixing mode. Signals falling beyond the preselector are rejected. This is the preselector in Figure 5.14.

Let us say that we are looking at an input range of 4 to 6 GHz, and we wish to somehow work exclusively on the 1– line. When the LO sweeps past 4.4 GHz (this is the frequency where it mixes with the 4.7 GHz input signal to produce the spurious IF signal), the preselector tunes its center frequency to 4.1 GHz and rejects whatever is present around 4.7 GHz. This 4.7 GHz now doesn't reach the mixer, and hence does not cause any spurious response on the display. Now when the LO sweeps past 5 GHz, the preselector allows the 4.7 GHz signal to reach the mixer, and we see the appropriate response on the display. The 5.3 GHz image signal is now rejected by the preselector, so it creates no mixing product and causes a wrong display. Finally, as the LO sweeps past 5.6 GHz, the preselector tunes itself to allow the 5.3 GHz signal to reach the mixer, and hence is properly displayed. The intersection of the mixing modes is totally eliminated, and with a narrow preselector BW (typically 35–80 MHz), it greatly attenuates all mixing and spurious responses. These filters also have rejections in the order of 70–80 dB. Since YIG resonators make up the preselector (discussed in Chapter 2), it does not work well at the low band. As observed, however, no mixing mode overlaps the 1– frequency mode in the low-band, high-IF case—so, that band is already taken care of.

5.6.4.2 External Harmonic Mixing

In the previous section we discussed internal harmonic mixing to tune the analyzer to the highest-frequency component mentioned in the data sheet. If we want to use this analyzer to test outside the maximum range, we employ a technique called external harmonic mixing (the external mixer output shown in Figure 5.14). This essentially uses an external mixer connected to the LO output of the first LO frequency. Here an LO_{OUT} port routes this signal to the mixer, which generates the required IF by mixing the input with some higher-order harmonic of the LO signal. This IF signal is then routed back to the IF in the port of the analyzer and then processed in the usual way for consequent display. As long as the IFs of both the harmonic mixer and the spectrum analyzer match, the signal can be processed and displayed internally. External mixers are available commercially from 18 to 325 GHz in selected bands. The harmonic mixing issues leading to spurious displays discussed in the previous section unfortunately present themselves in this case too; now it is severe, since we no longer have a preselector. The LO and its harmonics mix with the usual RF signal as well as some spurious tones present at the input. These in turn would lead to corresponding false displays after IF processing. To mitigate this, a preselector may be provided inside the harmonic mixer. However, at high frequencies, this filter may not perform satisfactorily. This introduces another technique, called signal identification, to overcome the previous drawbacks. The signal identification technique is classified into the image shift and the image suppress methods. This is an advanced topic and is discussed in Agilent (2004).

5.7 Dynamic Range and Sensitivity

One of the primary purposes of using a spectrum analyzer is to determine low-amplitude signals obscured by noise. This noise is internal to the spectrum analyzer and is a direct result of the usual random motion of electrons in all the devices present within it. This in turn is amplified by the gain stages and is displayed as a constant noise level on the spectrum analyzer, termed displayed average noise level (DANL).

When a standard 50 Ω load is used to terminate the spectrum analyzer, the equivalent noise generated in a 1 Hz bandwidth at room temperature is given by $kT_oB = -174$ dBm, where k = Boltzmann's constant (1.38×10^{-23} joule/K), T = temperature, in degrees Kelvin, conventionally assumed to be 290, and B = bandwidth in which the noise is measured, in Hertz.

Thus, the noise power is defined as -174 dBm/Hz at this temperature. This is amplified by the first gain stages and added to the internal noise of the same amplifier. The noise now is boosted to a considerable level, and the noise power added in subsequent stages is not that significant. Before the first gain stage, there are also attenuators present as discussed previously (the mixers also attenuate the signal and add their own noise). The DANL is also alternatively termed the noise floor of the spectrum analyzer. This is an effective number (not necessarily the true noise power, which has contributions from different stages of the instrument), as the analyzer display is calibrated to reflect the input signal power. Sometimes, we actually need to measure the noise of the input signal. This is easy to measure if the input noise is greater than the effective noise floor. One way of boosting the input signal is by reducing the RF attenuator value, such that the effective signal-to-noise ratio (SNR) goes up. In modern spectrum analyzers an internal microprocessor modifies the IF gain to compensate for the

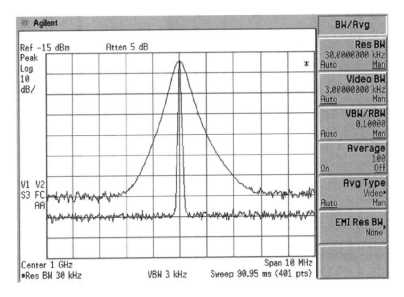

FIGURE 5.19
Displayed noise levels change as $10 \log (BW_2/BW_1)$, i.e., actual RBW. (c) Copyright Agilent Technologies 2004-5. Reproduced with permission.

changes in the levels of the input attenuator. This keeps the signals at the input of the analyzer constant while changing the displayed position of the noise. Another parameter that affects the signal-to-noise ratio at the output is the resolution bandwidth (RBW). Since the RBW filter appears after the IF gain amplifier, the bandwidth of this filter is critical in determining the bandwidth of the noise displayed. For example, a change in the resolution bandwidth by a factor of 10 changes the DANL by 10 dB (Figure 5.19). Alternatively, the video filter may also be used to smooth the variations of the amplitude of the input signal, which includes noise without affecting the constant signals. Since the video filter does not affect the DANL (note that by definition this is average level, so further averaging by the video filter is not significant), it also has no effect on the sensitivity of the analyzer. Hence, the best sensitivity for narrowband signals is obtained by selecting a very low RBW and minimum input attenuation. This, of course, leads to a greater sweep time.

5.7.1 Noise Figure

Another alternative characterization metric of a receiver is the noise figure. A standard definition of noise figure is

$$NF = \frac{SNR_{input}}{SNR_{output}} \tag{5.7}$$

Since the gain of the receiver is assumed to be 1 in our discussion, as the input signal amplitude is considered equal to the output displayed amplitude, this expression is simplified to

$$NF = \frac{N_O}{N_i} \qquad (5.8)$$

In logarithmic terms it is expressed in dB. To determine the noise figure of our spectrum analyzer, we simply need to measure the displayed noise in some BW, recalculate the numbers in a 1 Hz BW, and subtract it from the 50 Ω noise at room temperature in a 1 Hz BW (known to be –174 dBm). For example, say, we measure –110 dBm in a 10 KHz RBW. We first normalize this number to a 1 Hz BW. The normalized noise level is now –150 dBm. (Recall our earlier relation that a change in the resolution bandwidth by 10 changes the DANL by 10 dB.) So, $NF = 24$ dB. This noise figure (NF) is also independent of the RBW. Physically, a 24 dB noise figure means that an incoming signal has to be 24 dB greater than –174 dBm (when measured in a 1 Hz bandwidth), such that it comes up to the displayed average noise level of the analyzer. Recall that we saw in Chapter 2 that periodic signals have impulses in their spectra, and hence bandwidth is irrelevant for them, but for random signals (including modulated ones), measurement bandwidth will typically strongly affect the display.

5.7.2 Preamplifiers

A 24 dB NF (this is a realistic value) is too high for a receiver. In such cases very often a preamplifier with a high gain is cascaded in front of the analyzer to reduce the system noise figure and increase its sensitivity. However, a preamplifier also amplifies noise, and it is possible that the total input noise is now greater than the original input noise. Let us examine two separate cases.

The Friis formula for cascaded systems tells us that the overall noise figure for the preamplifier and spectrum analyzer is given by

$$F = F_{pre} + \frac{F_{spec} - 1}{G_{pre}} \qquad (5.9)$$

Here F is the overall noise figure, F_{pre} is the noise figure of the preamplfier, F_{spec} is the spectrum analyzer noise figure, and G_{pre} is the preamplifier available power gain. These numbers are *not* in dB.

Looking at realistic numbers, let us say that $F_{spec} = 200$ (23 dB), $F_{pre} = 4$ (6 dB), and $G_{pre} = 40$ (16 dB). Then, $F = 8.975$ (9.5 dB). It appears that we have achieved a great increase in the performance.

Now let us understand how this preamplifier affects our sensitivity—we will fix a 1 Hz bandwidth. Our DANL is now –174 + 9.5, or –164.5 dBm,

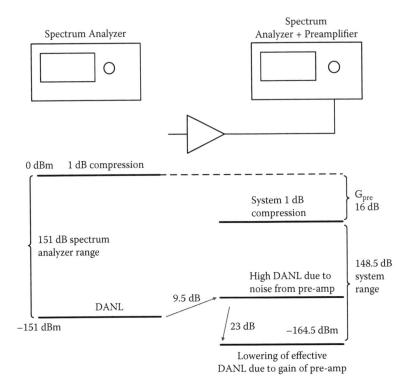

FIGURE 5.20
Decrease in measurement range when the preamplifier connected pulls up the noise floor.

while earlier it was –174 + 23, or –151 dBm. But the price is paid in terms of dynamic range. Suppose that without the preamplifier the maximum input power that could be correctly displayed (without gain compression—to be discussed soon) was 0 dBm. So, the dynamic range was 151 dB. Now the input to the preamplifier can at most be –16 dBm. This gives a dynamic range of 148.5 dB. So, we have lost out slightly in dynamic range. This effect is clarified in Figure 5.20.

5.8 Component Characterization

The obvious use of the spectrum analyzer is to measure the exact frequency and power in an unknown signal (or the various frequencies and associated powers). But this immediately leads to several component characterization procedures that are regularly carried out, specially for nonlinear components.

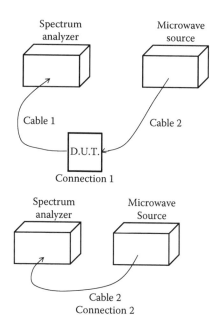

FIGURE 5.21
Component characterization using a spectrum analyzer.

The metrics primarily used to quantify measurements on nonlinear components are

1. Harmonic distortion
2. Intermodulation distortion
3. Gain compression
4. AM/PM (phase modulation) distortion
5. ACPR measurements

The simple setup for carrying out all these measurements is shown in Figure 5.21.

The approach is intuitively obvious, but here are some simple precautions to be taken. Let us suppose that the various powers (in dBm) and cable losses (in dB) are:

P_s: Power setting of the source
P_1: Power displayed by the spectrum analyzer in connection 1
P_2: Power displayed by the spectrum analyzer for connection 2
A_1: Attenuation of cable 1
A_2: Attenuation of cable 2

The attenuations are, of course, assumed to have been earlier measured using a network analyzer.

It is very important to realize that other than possibly P_s, the other four quantities are *functions of frequency* (possibly a few discrete frequency points). If the source frequency is changed to obtain readings at many input (fundamental) frequencies, then P_1 and P_2 values will be observed for many harmonics of each of these input frequencies. For some measurements (e.g., intermodulation), even the source will provide power at multiple frequencies simultaneously. In such a case it is understood that what is shown as source in Figure 5.21 actually consists of multiple sources, with a power combiner included.

Assuming a single-source frequency, the power available (Ha, 1981) from the source $P_{av} = P_2$, and also $P_{av} = Ps - A_2$. If these two values are not close, the instruments should be checked for calibration. For a small discrepancy (typically < 0.5 dB), or if the spectrum analyzer is known to be well calibrated, P_2 is taken to be the correct value. The power delivered (Ha, 1981) to the load (50 Ω—the spectrum analyzer input) $= P_1 + A_1$. Again, these numbers have to be observed at multiple frequency points. If adequate power is available, it is desirable to insert an attenuator (10/20/40 dB broadband coaxial attenuators are common). This ensures that the DUT is loaded by exactly 50 Ω. For oscillator testing and surveillance type of measurements, there is of course no microwave source. Now only connection 1 of Figure 5.1, that too without the source and cable 2, is used.

From these powers (available and delivered), components may be characterized in different ways; some examples are given subsequently.

5.8.1 Harmonic Distortion

As is well known, any nonlinear component generates harmonic products of the fundamental signal. To measure these products accurately, it is essential that they are clearly differentiated from the internal distortion products generated by the mixer inside the spectrum analyzer. Several methods have already been discussed earlier. Let us now look into the fundamental behavior of a harmonic product. When a sinusoid is applied to a nonlinear system, the output generally consists of the original signal and components that are integer multiples of the fundamental frequency. If $x(t) = A \cos(\omega t)$, then limiting ourselves to the maximum third-order nonlinearity yields the following:

$$y(t) = \alpha_1 A \cos(\omega t) + \alpha_2 A^2 \cos^2 \omega t + \alpha_3 A^3 \cos^3 \omega t$$

$$= \alpha_1 A \cos(\omega t) + \frac{\alpha_2 A^2}{2}(1 + \cos 2\omega t) + \frac{\alpha_3 A^3}{4}(3\cos \omega t + \cos 3\omega t)$$

$$= \frac{\alpha_2 A^2}{2} + \left(\alpha_1 A + \frac{3\alpha_3 A^3}{4}\right)\cos\omega t + \frac{\alpha_2 A^2}{2}\cos 2\omega t + \frac{\alpha_3 A^3}{4}\cos 3\omega t$$

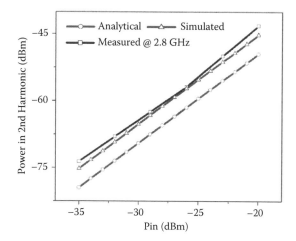

FIGURE 5.22
Measured, simulated, and modeled second harmonic at 2.8 GHz.

If we now look into the above carefully, we see that the first term is a dc term. This arises from second-order nonlinearity, often called a dc offset. The second term is the fundamental, the third is the second harmonic, and the last term is the third harmonic. The even-order harmonic terms may disappear completely if the system has an odd symmetry, or in other words, when it is fully differential in nature. However, a complete cancelation of this is not possible because of the random mismatch among transistors in the differential pair. Harmonic distortion is not that important in narrowband systems for obvious reasons. However, to cater to modern wireless standards, most of the products today are broadband. In such cases, where maybe decade or octave wide operation is encouraged, the harmonic of one frequency at the lower edge of the band may go and interfere with the actual signal at the middle or upper edge of the band. Figures 5.22 and 5.23 show the actual measured second and third harmonic powers of a complementary metal oxide semiconductor (CMOS) LNA at 2.8 GHz (input, only fundamental present) when the input power is varied. The second harmonic goes up 2 dB/dB for every 1 dB rise in input power, while the third harmonic product usually has a slope of 3:1. Beyond –20 dBm input, all these curves enter compression (discussed later).

5.8.2 Third-Order Intermodulation Distortion

Similar to the harmonics of a single-tone CW signal, a nonlinear component generates intermodulation products when excited by a two-tone signal. This again has to be differentiated by similar means from the intermodulation products generated internally by the mixer. Let us now look at this in some

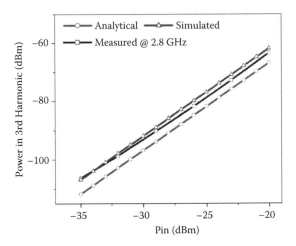

FIGURE 5.23
Measured, simulated, and modeled third harmonic at 2.8 GHz.

detail. Let us assume that we have two interferers ω_1 and ω_2, and these are applied to our nonlinear system. The output now also contains undesired higher-order products along with the harmonics of the fundamental signals. This is actually termed intermodulation as a result of the mixing of these two components as their sum is raised to powers greater than unity. Assuming $x(t) = A_1 \cos(\omega_1 t) + A_2 \cos(\omega_2 t)$, then

$$y(t) = \alpha_1 (A_1 \cos(\omega_1 t) + A_2 \cos(\omega_2 t)) + \alpha_2 (A_1 \cos(\omega_1 t) + A_2 \cos(\omega_2 t))^2 + \alpha_3 (A_1 \cos(\omega_1 t) + A_2 \cos(\omega_2 t))^3$$

Now if we discard the dc, fundamental, and harmonic terms in the above expression, we are left with the intermodulation product terms at

$$\omega = 2\omega_1 \pm \omega_2 : \frac{3\alpha_3 A_1^2 A_2}{4} \cos(2\omega_1 + \omega_2)t + \frac{3\alpha_3 A_1^2 A_2}{4} \cos(2\omega_1 - \omega_2)t$$

$$\omega = 2\omega_2 \pm \omega_1 : \frac{3\alpha_3 A_1 A_2^2}{4} \cos(2\omega_2 + \omega_1)t + \frac{3\alpha_3 A_2^2 A_1}{4} \cos(2\omega_2 - \omega_1)t$$

This is illustrated in Figure 5.24.

Let us say that an antenna is tuned to receive a signal at ω_0. Unfortunately, there are also two large interferers present at ω_1 and ω_2. This is now fed to a low-noise amplifier. Also assume that the product $2\omega_1 - \omega_2$ falls at ω_0. In such a case, the desired smaller-amplitude signal is totally corrupted by the large interferer. Figure 5.25 illustrates a practical case, when two tones at 2.8 and 2.825 GHz are added and given as input to a 2.8 GHz CMOS LNA. The

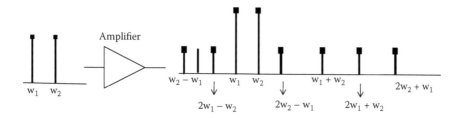

FIGURE 5.24
Generation of various intermodulation products from a two-tone input signal.

resultant third-order intermodulation product has a slope of 3:1 and ultimately (if extrapolated) intersects the fundamental tone curve at a very high theoretical input power (beyond the measurement range of course), termed the third-order input intercept point.

5.8.3 Gain Compression

In the measurements of a nonlinear component, gain compression is an important phenomenon. This may be defined as the condition when, for every 1 dB rise in input power, the output does not raise by 1 dB. When the level of the input signal rises, it becomes high enough such that the output is no longer a linear function of the input. The component goes into saturation and the displayed signal is low. The phenomenon is important in mixers

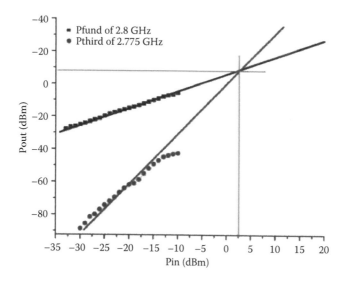

FIGURE 5.25
Intermodulation in a CMOS LNA.

where the input (RF) and output (IF) are *at different frequencies*. Typically this gain compression happens for a mixer input power of –5 to +5 dBm (for a passive mixer, gain < 1 or gain < 0 dB). This helps us to choose the appropriate RF attenuator settings such that the amplitude of the input signal reaching the mixer is low, and hence it operates in the linear region. The spectrum analyzer itself uses mixers, as explained earlier, and in case the amplitude of the incoming signal is high and is outside the ADC's range, analyzers with digital IF show an ADC overload error message.

In general, for microwave components, there are different situations where this compression is measured, some being:

1. **CW compression:** This is the most conventional method where the change in gain in a device is measured when the input power is swept linearly upward.

2. **Mixer two-tone compression:** This measures the change in small signal gain of a mixer when the large signal (LO) power is swept upward.

3. **Pulsed power compression:** This measures the change in system gain to a narrowband or broadband RF pulse when its power is swept upward. While completing pulse-based measurements, we use a pulse width far broader than the RBW of the filter. This displays the pulse level well below the actual input pulse level. This may lead us to being unaware that the actual signal is above the mixer compression threshold. Use of attenuators (or increasing the attenuation setting in the spectrum analyzer) will bring the power to acceptable levels. However, if there are high-power and low-power signals coexisting, which exceed the dynamic range of the instrument, then additional components will be required—basically an application-specific measurement setup will have to be devised.

A practical measured gain compression scenario in a CMOS RF power amplifier is demonstrated in Figure 5.26. Note how the output power curve bends after a certain level of input power announcing the onset of gain compression.

5.8.4 AM/PM Distortion

In nonlinear RF circuits (for example, power amplifiers) the amplitude modulation is converted to phase modulation, thus creating undesired effects. This is broadly defined as the dependence of the phase shift on the input signal amplitude. For an input signal of $V_{in}(t) = V_1 \cos\omega_1 t$, the fundamental output component is given by

$$V_{OUT}(t) = V_2 \cos(\omega_1 t + \phi(V_1))$$

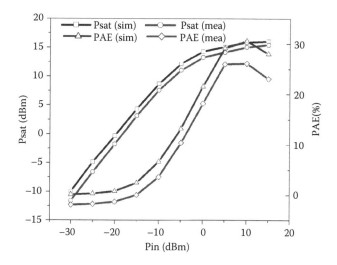

FIGURE 5.26
Gain compression in a CMOS RF power amplifier.

This $\phi(V_1)$ is the input-dependent phase shift in the output signal. Fundamentally, it is the change in phase of large signal gain of an amplifier under higher levels of excitation. Ordinary spectrum analyzers will not be able to measure this phenomenon—the vector signal analyzers can used for this measurement, or even a vector network analyzer can be used, possibly with an amplifier at the component input to boost the signal available and an attenuator at the output to avoid overloading the vector network analyzer (VNA).

5.9 Conclusion

An overview of the functioning of a spectrum analyzer has been given. Also, some applications of this instrument have been described. The closely related vector signal analyzer has been mentioned, but it has not been discussed here—a detailed description of this enhanced spectrum analyzer can be found in the literature (particularly instrument data sheets).

Problems

1. A spectrum analyzer is used to analyze the power distribution at various frequencies in an AM signal with carrier 1 MHz and an audio message signal (20–20,000 Hz). What RBW values can be used?

2. The RBW filter of a properly calibrated spectrum analyzer with 50 Ω input has $|S_{21}|$ as shown:

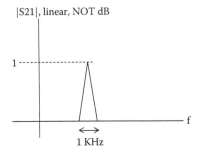

What is the display if the input is a voltage source (with a 50 π series resistance):

$$V(t) = 1.0 \cos(2\pi\ 10^6 t) + 0.5 \cos(2\pi\ (1000.2)10^3\ t)$$

3. If a wide-band tunable preselector can be constructed, what is the need for a spectrum analyzer at all? This filter, coupled with a power meter, will display the power spectral density without complications like images and multiple responses.

4. Assume that a voice signal (through a microphone) is connected to the above spectrum analyzer (same RBW). What will be the displayed power at 1.6 KHz if the power spectral density (single-sided) of the microphone output has a Gaussian shape, with peak value = 10^{-6} W/Hz, mean = 1.6 KHz, and standard deviation = 1.5 KHz? Assume that the envelope detector records the true power (integral of the PSD over the RBW) of its input signal. What would be the peak power displayed if the microphone output is multiplied by $\cos(2\pi\ 10^6 t)$?

5. Redesign the configuration in Figure 5.14 to enable measurement inputs up to 6 GHz *without* using any tunable filter. The LO has a range of 3–7 GHz, and fixed filters of reasonable specifications can be used.

References

Agilent. *Spectrum Analysis Basics*. Application Note 150. Agilent Technologies (2004).
Ha, T.T. *Solid State Microwave Amplifier Design*. John Wiley & Sons, New York, 1981.

6

Noise Measurements

Noise figure is one of the important metrics to assess the performance of an amplifier or mixer that appears as the first component in a receiver. The terms *noise figure* and *noise factor* are used interchangeably in the literature. However, strictly speaking, noise figure (NF) is defined as the decibel version of the noise factor F. This is applicable to additive noise—we will discuss phase noise later.

6.1 Definition

Noise factor (F) is the factor by which an amplifier degrades the signal-to-noise ratio (SNR) of the input signal. It is never smaller than unity.

$$F = \frac{SNR_i}{SNR_o}, \quad NF = 10\log_{10}\left(\frac{SNR_i}{SNR_o}\right) \tag{6.1}$$

Here SNR_i = signal power-to-noise power ratio at the input of the component, while SNR_o is the corresponding quantity at the output. All powers are assumed to be available powers—these would be the powers dissipated *if a conjugate-matched load is used*; otherwise, the actual power dissipated will be less.

The temperature at which the noise measurements are made is of concern. Temperature of the source has a profound impact on the measurements. If the source temperature is very low, then the corresponding source noise is low. In such a case the self-generated noise added by the device under test (DUT) will have a comparatively greater effect and the measured noise figure will be higher than if the source were hotter. Because of this sensitivity, a standard has been adopted: noise figure measurements are made at a standard temperature T_0 of 290 K (= 17°C). This means that the noise power available at the input is always kT_0B, where B is the bandwidth of the system and k (Boltzmann's constant) = 1.38×10^{-23} J/K. More clarity about the bandwidth B will emerge shortly.

6.2 Noise Measurement Basics

Noise figure measurements must be made with a source that is a matched termination at 290 K. In co-axial systems this is invariably a 50 Ω resistor. In waveguides it is a waveguide load (which gives negligible reflections when connected to a standard waveguide for the concerned band). One should be clear with the definition of available power in this case, as discussed above. In case the load is the input impedance of an amplifier, it will be frequently not matched—in this case the actual power going into the amplifier will be less than the available power.

With T_0 fixed, the noise figure is an intrinsic property of the device alone. In actual measurement the noise appearing at the output of DUT is the sum of amplified available source noise power at $T_0 = 290$ K (N_{os}) and that generated by the DUT (N_a) itself. This means that we are restricting the noise to the additive type. A different type (phase noise) will be considered later.

$$N_{out} = N_{os} + N_a = kT_0 B G_{av} + N_a \tag{6.2}$$

where B is the bandwidth and G_{av} is the available power gain (Ha, 1981) of the DUT. If required, G_{av} can be calculated from the S-parameters using:

$$G_{av} = \frac{\left(1 - |\Gamma_s|^2\right)|S_{21}|^2}{|1 - S_{11}\Gamma_s|^2 \left(1 - |\Gamma_{out}|^2\right)}$$

where

$$\Gamma_s = \frac{Z_s - Z_0}{Z_s + Z_0}$$

and

$$\Gamma_{out} = S_{22} + \frac{S_{21}S_{12}}{1 - S_{11}\Gamma_s}$$

Here Z_s is the source impedance and Z_0 is the reference for S-parameters (usually 50 Ω). In most cases, $|S_{21}|^2$ is substituted for G_{av} since Γ_s and Γ_{out} are usually small for well-designed systems.

From the definition of noise factor,

$$F = \frac{SNR_i}{SNR_o} = \frac{Si/Ni}{So/No} = \left(\frac{Si}{So}\right)\left(\frac{No}{Ni}\right)$$

Substituting G_{av} for the ratio of output signal to the input signal, kT_0B, for available input noise power in Equation (6.2), we may write

$$F = \frac{kT_0BG_{av} + N_a}{kT_0BG_{av}} = 1 + \frac{N_a}{kT_0BG_{av}} \quad (6.3)$$

Noise figure determination using the above equation requires the measurement of noise added, available gain, and bandwidth. However, it is not easy to measure N_a, and the product of effective bandwidth and available gain BG_{av} accurately.

By performing the noise measurements at two different temperatures, one can do away with the need to measure the gain-bandwidth product directly. As noise factor is a dimensionless quantity, these ratiometric measurements are possible and advantageous.

As shown in Figure 6.1, measurements are done at two different temperatures (hot and cold). In the simplest case, one with the source at room temperature and the other at a higher temperature will suffice. The use of a hot source increases the component of output noise without changing the noise added by the DUT. If the source temperatures are accurately known, one can solve for the noise added by DUT, and thereby the noise figure.

Two points on the plot are enough to determine the line, thus eliminating the need to measure BG_{av}—the slope. This graph is the basis of most noise figure measurements.

From the above plot, Equation (6.3) for noise factor can be written as

$$F = 1 + \frac{N_a}{kT_0BG_{av}} = 1 + \frac{y - intercept}{(T_0)(slope)} \quad (6.4)$$

If we make noise measurements at a high temperature T_h that is above T_0 by a value T_{ex}, and another at T_c, then the output noise powers are given by

$$N_1 = KT_cBG_{av} + N_a, \quad and \quad N_2 = KT_hBG_{av} + N_a \quad (6.5)$$

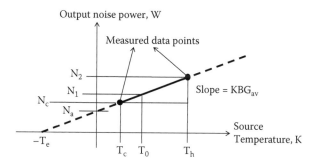

FIGURE 6.1
Plot of output noise power as a function of source temperature.

Subtracting these, we get

$$KBG_{av} = \frac{N_2 - N_1}{T_h - T_c}$$

Substituting this expression for KBG_{av} in the second part of (6.5) we get

$$N_a = N_2 - T_h \frac{N_2 - N_1}{T_h - T_c} = \frac{N_1 T_h - N_2 T_c}{T_h - T_c}$$

Using these expressions for KBG_{av} and N_a in (6.5), we get

$$F = \frac{\dfrac{T_{ex}}{T_0} - Y\left(\dfrac{T_c}{T_0} - 1\right)}{Y - 1} \tag{6.6}$$

Here, $Y = N_2/N_1$ and $T_{ex} = T_h - T_0$.

The ratio $\frac{T_h - T_0}{T_0} = \frac{T_{ex}}{T_0}$ is called the excess noise ratio, which is generally supplied by the manufacturer for a given diode-based noise source (described below). However, it is precisely valid if the noise source impedance is the same as the specified standard impedance (generally 50 Ω), as the *available powers* are considered in our definitions. For a termination actually kept at different temperatures, it is not an issue. Finally, everything boils down to an accurate measurement of the output noise power ratio Y—this approach is hence referred to as the Y-factor method (Agilent, 2010).

Traditionally, actual hot and cold sources with resistors at the boiling point of water (373 K) and at the boiling point of nitrogen (77 K) were used. The greater the temperature difference, the better the results are, as the slope and intercept can be calculated accurately for a given uncertainty in the measurements.

At higher temperatures it is difficult to make direct measurements. Measurements using noise diodes that can generate the same noise as an extremely hot termination while the diodes operate at room temperature are performed nowadays (effectively, the noise generated is the same as a hypothetical termination kept at >1000 K). By turning off the diode, it acts as a cold source with its internal resistance network providing the noise power at the ambient temperature. However, unlike hot and cold resistors, its noise cannot be computed from first principles. The circuit of a typical noise source using a diode is shown in Figure 6.2.

An alternative figure of merit that estimates the noise performance of the amplifier is noise temperature (T_e). From Figure 6.1 noise temperature is the extrapolated intercept of the noise power line with temperature axis, with reversed sign. This means that

$$N_a = KT_e BG_{av} \Rightarrow N_{out} = k(T_s + T_e)BG_{av} \tag{6.7}$$

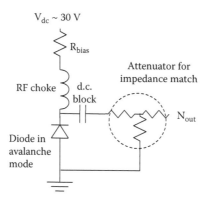

$V_{dc} \sim 30\ V$

R_{bias}

RF choke   d.c.
            block

Attenuator for
impedance match

N_{out}

Diode in
avalanche
mode

FIGURE 6.2
Use of a noise diode.

where T_s is the source temperature. From (6.7) noise temperature can be interpreted in the following way. If the component under consideration is somehow made noiseless (i.e., it adds no noise), then we must raise the temperature of termination at the input, if the output noise is to be brought up to the level that is observed with the real noisy component. This rise in temperature is T_e.

From (6.3) this means that

$$F = 1 + \frac{T_e}{T_0}, \quad or \quad T_e = (F-1)T_0 \tag{6.8}$$

Some example values of noise figure (NF) and noise temperature (T_e) are shown in Table 6.1.

In general, noise temperatures are easier to relate to, compared to noise figures, especially at low values. Of course, both convey the same information. Combining (6.6) and (6.8) we get

$$T_e = \frac{T_h - YT_c}{Y-1} \tag{6.9}$$

TABLE 6.1

Conversion of Noise Figure and Noise Temperature

Noise Figure	Noise Temperature
0.3 dB	20.7
0.35 dB	24.3
10 dB	2610
11 dB	3361

One major conceptual difference between noise figure and noise temperature is that noise temperature does not require any convention regarding a standardized T_0; it is purely dependent on the system, assuming that added noise from the DUT is proportional to the bandwidth. This point is discussed next.

6.2.1 Spot Noise Figure

This is measured in a narrowband centered around a frequency. It is known that the inherent additive noise inside active or lossy devices varies only slowly with frequency—of course, sharp variations in G_{av} may certainly change the noise appearing at the output. If the bandwidth is small (roughly the frequency range over which G_{av} does not change significantly), then N_a is proportional to bandwidth. In this case, by virtue of (6.3), F is independent of B.

Now the measurement bandwidth is fixed at a certain value (it does not matter how much), and the noise figure is reported at different frequencies over a wide band.

On the other hand, (6.3) or (6.6) may be used to calculate the noise figure of, say, a 2–18 GHz amplifier, taking B = 16 GHz. Interpretation of such a result may be difficult, as there is likely to be significant variation of gain and impedances over such a large bandwidth.

6.2.2 Noise Figure of Cascaded Systems

While computing the noise figure of a cascaded system, one should keep in mind that each stage sees a different source impedance dependent on the previous stage, and hence the noise figure must be computed with respect to that source impedance for a given stage.

Consider a two-stage cascaded system as shown in Figure 6.3. G_1 and F_1 are the available power gain and noise figures of DUT1, and G_2 and F_2 are for DUT2. G_2 is measured with the output impedance of the first stage as its source impedance in accordance with the definition of available gain (Ha, 1981).

Output noise power (N_{o1}) at the end of the first stage is given by

$$N_{01} = kT_s BG_{av1} + kT_{e1} BG_{av1} = k(T_s + T_{e1}) BG_{av1}$$

FIGURE 6.3
Noise from cascaded components.

This noise is further amplified as it passes through the second stage and appears at the output along with the inherent noise generated by the second stage ($N_{a2} = kT_{e2}BG_{av2}$):

$$N_{02} = k(T_s + T_{e1})BG_{av1}G_{av2} + kT_{e2}BG_{av2}$$

that is,

$$N_{02} = k\left(T_s + T_{e1} + \frac{T_{e2}}{G_{av1}}\right)BG_{av1}G_{av2}$$

Here we have assumed that $N_{a2} = kT_{e2}BG_{av2}$ irrespective of the impedance seen by the input of the second block (i.e., the output impedance of the first block), so T_{e2} is not dependent on the output impedance of the first block. To be rigorous, this is not correct, and a much more involved calculation incorporating the variation in added noise with source impedance has to be undertaken (Ha, 1981). In practice, this is rarely required.

Looking at (6.7), the noise temperature of the cascade is given by

$$T_{e12} = T_{e1} + \frac{T_{e2}}{G_{av1}} \tag{6.10}$$

If we extend the above equation for more stages, we have the overall noise figure as

$$T_e = T_{e1} + \frac{T_{e2}}{G_{av1}} + \frac{T_{e3}}{G_{av1}G_{av2}} + \ldots \tag{6.11}$$

Substituting noise figures for the corresponding noise temperatures, this can be further written as (Friis formula)

$$F_{12} = F_1 + \frac{F_2 - 1}{G_{av1}} + \frac{F_3 - 1}{G_{av1}G_{av2}} + \ldots \tag{6.12}$$

At this stage we are in a position to outline a simple procedure for measuring noise figure:

1. Arrange the setup as in Figure 6.4. Notice that the gain of the spectrum analyzer is assumed to be 1, since if input power = 1 mW, it displays 1 mW (proper calibration is assumed). Actually, G_3 is not required, looking at (6.12).
2. Measure F of the whole system (three cascaded blocks) using (6.6).
3. Use the Friis formula (6.12) to calculate F_1, since all other terms are known.

FIGURE 6.4
Simple setup for measuring noise figure.

As usual, the gains are, strictly speaking, available gain values. In practice, substituting $|S_{21}|^2$ for G gives results that are acceptable except in critical applications (e.g., ultra-low-noise amplifiers for radio astronomy). If required, the true G_1 and G_2 can be calculated—since S-parameters of the DUT and pre-amp can be accurately measured with a vector network analyzer, and the input impedance of the spectrum analyzer can be taken to be 50 Ω.

If F_2 is not known, then this has to be determined first. For this, the DUT should be omitted, and the termination should be connected at the input of the pre-amp in Figure 6.4. Following the same procedure outlines above, F_2 can be evaluated using the Friis formula for two cascaded systems.

If even F_3 is not known, this also has to be evaluated. For this, the termination at T_0 should be connected to the spectrum analyzer input, and the displayed noise measured (to minimize spectrum analyzer noise, 1 or 10 Hz is the recommended RBW setting for the entire measurement). This gives N_a for the instrument, and hence F_3 is determined (B = RBW and G = 1 in (6.3)). In most spectrum analyzers, F_3 is very large (>20 dB), so N_a for the spectrum analyzer tends to be large, and it does not really matter if the termination is at T_0 or room temperature—the displayed noise is almost entirely N_a.

The spectrum analyzer is not always a good choice for noise measurements. It has a high noise figure, and it may not be easy to arrive at an unambiguous value for the displayed noise power. For better accuracy, a diode-based power sensor, preceded by an amplifier and a band-pass filter (to decide B) and followed by an oscilloscope, works well. The amplifier is, of course, already shown in Figure 6.4 (the preamplifier). If $(G_1 \, G_2)$ is large, then (6.11) tells us that we need not worry about additional noise from the diode sensor.

6.3 Special Consideration for Mixers

For mixers, single-sideband (SSB) or double-sideband (DSB) noise figure measurements may be performed. This complication results due to the image effect in a mixer, as shown in Figure 6.5.

Suppose that the RF signal occupies the band f_1 to f_2, while the mixer LO is slightly below at f_L. The band f_1 to f_2 then gets translated to $(f_1 - f_L, f_2 - f_L)$:

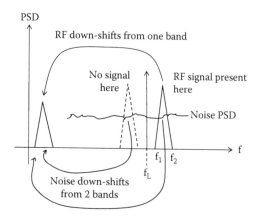

FIGURE 6.5
Effect of mixer on signal and noise.

a much lower band, the IF. But there is also another band of frequencies that can mix with f_L and give the same band (IF). These frequencies are $(f_L - f_2, f_L - f_1)$, shown by the dashed lines in Figure 6.4. This is called the image of the RF band. Since there is no signal here, it does not contribute to the IF signal power. However, noise, being broadband, is present in the image band as well. In case the separation between the RF and image band edges (this is $2(f_1 - f_L)$) is large enough to permit removal of noise at the input with a band-pass filter, this should be done. Unfortunately, for most cases (e.g., $f_1 = 900$ MHz and $f_L = 899.9$ MHz are typical numbers) the sharpness of the filter cutoff is impractical. Now, regarding the mixer as a two-port linear component, with RF input and IF output, let us evaluate the noise figure. Note that the LO is now assumed to be an integral part of the mixer. Its noise will contribute to the N_a of the mixer.

Suppose that the input signal power is S, which is concentrated in the RF band only.

Let the input noise power spectral density be $kT_0 = N_0$ (assumed single-sided, i.e., only defined for positive frequencies). Then input SNR is $\frac{S}{BN_0}$, where $B = f_2 - f_1$. For SNR calculation, the signal and noise have to obviously occupy the same band.

Now the mixer may have different gains when the input is in the RF band and when the input is in the image band. Let us call these G_R and G_I, respectively. Usually these are <1 (i.e., < 0 dB).

So, the output signal power is SG_R.

The output noise power is $N_0 B (G_R + G_I) + N_a$, since noise from both RF and image bands have been translated to the same IF band. Now we can express the noise factor (called SSB noise factor/figure):

$$F_{SSB} = \frac{S/(BN_0)}{SG_R/(N_0B(G_R+G_I)+N_a)} = \frac{N_0B(G_R+G_I)+N_a}{G_RBN_0} \tag{6.13}$$

Even for an ideal mixer with $N_a = 0$, this $[1 + G_I/G_R]$ is >1, and if $G_R = G_I$, the noise figure will be 3 dB. *This is correct and physically meaningful*; if G_I could be made 0, then indeed a mixer with no N_a will show $NF = 0$ dB. Such a mixer is called an image-reject mixer, and indeed this can be made (Maas, 1993). However, frequently it is not practical to implement an image-reject mixer, as these employ couplers and can be large in size.

Sometimes, to ensure that $NF = 0$ dB if $N_a = 0$, a different definition is used. Now the denominator in (6.13) is modified using the artificial argument: since noise is downconverted from both RF and image bands, the image noise should also be considered while calculating input SNR.

This leads to a different noise factor F_{DSB} (double-sideband noise factor/figure):

$$F_{DSB} = \frac{N_0 B(G_R + G_I) + N_a}{G_R B N_0 + G_I B N_0} \tag{6.14}$$

This is lower than F_{SSB} by a factor $(1 + G_I/G_R)$—number usually close to 2 (or 3 dB if noise figure is described).

In any case, F_{DSB} is of interest only because this is found in some data sheets, since being 3 dB smaller, it gives the impression of better noise performance.

Measurement of the SSB noise figure is carried out in exactly the way described above, with a few minor considerations:

1. A low-pass (or sometimes band-pass) filter is usually connected after the mixer, and the noise figure of this combination is measured.

2. In case the input of the mixer is in waveguide, the termination will no longer be 50 Ω, but a waveguide-matched load.

3. An upconverting mixer at the higher frequencies ($IF > 20$ GHz) may be difficult to characterize, but these usually give the same performance when the RF and IF ports are reversed. Of course, this is not true if the mixer integrates an IF or RF amplifier or an isolator.

6.4 Phase Noise

Until now we have considered additive noise. The other category of noise comprises amplitude modulation (AM) noise and phase noise. As discussed in Chapter 2, these take a mathematical form:

AM noise: $x(t) = A(t) \cos(2\pi f_0 t) = (A_0 + n(t)) \cos(2\pi f_0 t)$

Phase noise: $x(t) = A \cos(2\pi f_0 t + \phi(t))$

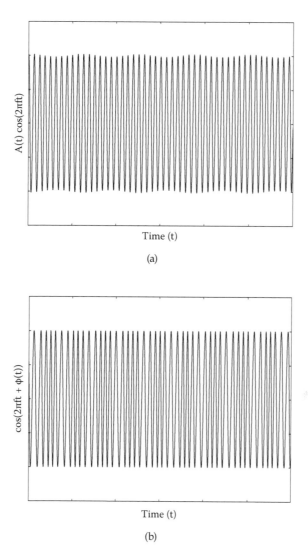

FIGURE 6.6
(a) AM noise. (b) Phase noise.

In general, these may even appear combined (i.e., $A(t) \cos(2\pi f_0 t + \phi(t))$), but amplitude fluctuations may be removed using a limiter or a zero-crossing detector, and it is phase noise that is more imprtant.

Waveforms illustrating such signals are shown in Figure 6.6. Let us focus on the phase noise.

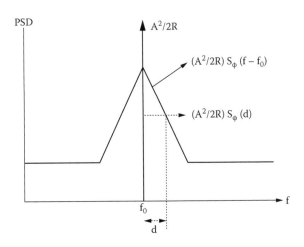

FIGURE 6.7
Phase noise spectrum.

From Chapter 2, the power spectral density of $x(t) = A \cos(2\pi f_0 t + \phi(t))$, in single-sided form, is

$$S_x(f) = \frac{A^2}{2R}[\delta(f - f_0)] + \frac{A^2}{2R}[S_\phi(f - f_0)] \tag{6.15}$$

This is plotted in Figure 6.7 for a hypothetical triangular shape of the power spectral density (PSD) of ϕ (only positive frequencies are considered, so values are doubled from expressions in Chapter 2). An important consideration is that although the plot is single-sided, $S_\phi(f)$ *is still the double-sided power spectral density of* $\phi(t)$. The unit of $S_\phi(f)$ is Hz^{-1} (i.e., seconds). Obviously there is no resistor R used to define $S_\phi(f)$. When displayed using a spectrum analyzer, we will get the plot shown in Figure 6.8—the RBW filter shape appears in place of the hypothetical impulse, as discussed in Chapter 5. This is acutally highly exaggerated, with linear vertical axis. In practice, spectrum analyzers are almost always used with a dBm scale.

For quantification of phase noise, IEEE standard 1139-1999 recommends the use of $S_\phi(f)$, the *double-sided* PSD of $\phi(t)$, but the offset is taken only on one side, i.e., d is strictly positive, and the value from $-d$ is *not* added. To put it in another way, suppose $S_{\phi 1}(f)$ is the single-sided PSD of $\phi(t)$, where f ranges from 0 to ∞ and power in negative frequencies is accounted for by doubling the power at the corresponding positive frequency. If $\phi(t)$ was a voltage signal, then the power dissipated by it in a 1 Ω resistor preceded by a 1 Hz

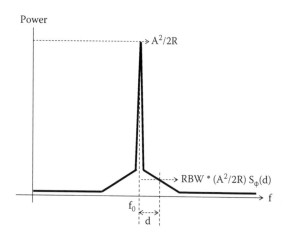

FIGURE 6.8
Spectrum analyzer display of phase noise.

bandwidth ideal filter centered at f_0 would actually be $S_{\phi 1}(f_0)$. Now $[S_{\phi 1}(f)/2]$ gives the phase noise.

Traditionally, information about $S_\phi(f)$ is usually conveyed through a few (typically three or four) points on the plot of $S_\phi(f)$, taken for different values of d (refer to Figure 6.7). For example, if the PSD at $d = 1$ KHz is 1 pW/Hz while the carrier power ($A^2/2R$) is 1 mW, then the phase noise is expressed as $[10^{-12}/10^{-3}]$, which in dB is -90 dB. This is commonly expressed by the phrase "phase noise at 1 KHz offset $= -90$ dBc/Hz." *dBc* stands for "dB with respect to carrier." Looking at Figure 6.7 again,

$$(A^2/2R) = 10^{-3}$$

while

$$(A^2/2R)\, S_\phi(f) = 10^{-12}$$

which means that $S_\phi(f) = 10^{-9}$, and $10 \log[S_\phi(f)] = -90$; i.e., the noise power in the 1 Hz bandwidth is -90 dB below the carrier power. This is the origin of the term *dBc*. Strictly speaking, this is not good terminology, since we are expressing a quantity ($S_\phi(f)$) that has units, in dB. However, these terms are nowadays standard. Also, for modern equipment or components, an actual plot of $S_\phi(f)$ is usually provided by manufacturers instead of a few points, since automated test equipment is commonly available today for measuring phase noise.

The same information can be extracted from a spectrum analyzer display as in Figure 6.8, using the formula

$$\mathcal{L}(d) = 10 \log \frac{(displayed\ power,\ offset\ from\ carrier = d)/RBW}{carrier\ power} \tag{6.16}$$

$\mathcal{L}$ is the symbol usually employed for phase noise as defined in (6.16). The origin of the term *dBc/Hz* is clear from this expression.

6.5 Phase Noise Measurement Techniques

6.5.1 Spectrum Analyzer Technique

In this technique the power spectral density is directly measured by connecting the device to a spectrum analyzer as shown in Figure 6.8. This is the simplest option; unfortunately, many oscillators available today have phase noise low enough to be masked by the noise floor (also known as DANL— see Chapter 5) of spectrum analyzers. Even when this is not the case, the technique can only be applied if the spectrum analyzer's internal source has a phase noise far less than the DUT—such is not always the case (say, a simple crystal oscillator is the DUT). This technique is therefore of limited use.

6.5.2 Phase Detector Techniques

In this the difference of phases in two signals is converted to a corresponding voltage output. The basic concept is shown in Figure 6.9. Double-balanced mixers, with both inputs strong enough to just saturate the mixer, are normally employed—this makes the measurement immune to AM noise.

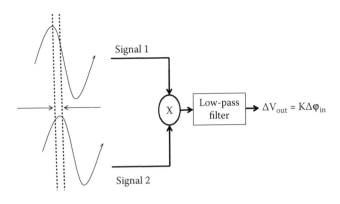

FIGURE 6.9
Basis of phase detector techniques.

If signal 1 is $A \cos(\Omega t + \theta(t))$ and signal 2 is $B \sin(\Omega t + \phi(t))$, then the low-pass-filtered output can be seen by taking the product to be $K \sin(\phi(t) - \theta(t)) \cong K (\phi(t) - \theta(t))$. This can be amplified and processed to yield relevant information. However, there are several problems that have to be addressed before this technique can be made practical:

1. We have assumed that Ω is the same for both. If we use two separate independent sources, these will have some phase difference that will fluctuate, leading to an additional phase noise that will obscure the result. So, there should be some form of synchronization of the phases.
2. We have assumed that the signals are in quadrature. This is, strictly speaking, not essential, but it will give the best sensitivity—for example, if the signals are in phase, then the output will be $K \cos(\phi(t) - \theta(t)) \cong K$ for small $\phi(t)$ and $\theta(t)$, which will not yield anything useful.

Let us see how these issues have been addressed in measurement systems.

6.5.3 Reference Source/PLL Method

A mixer is used as a phase detector in this method, with DUT connected to one of the inputs and a reference source that follows the DUT at the same carrier frequency and in phase quadrature with the DUT's output. The mixer sum is filtered by a low-pass filter, and the difference signal contains the ac voltage fluctuations proportional to the noise contributions from both sources. It is often possible to extract the DUT phase noise from this. The implementation is shown in Figure 6.10.

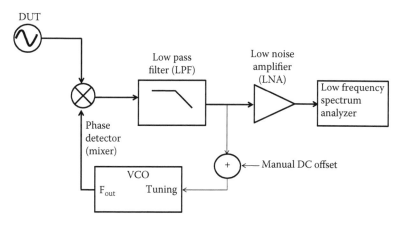

FIGURE 6.10
PLL technique for phase noise measurement.

Here, the voltage-controlled oscillator (VCO) generates signal 2 shown in Figure 6.9. The loop keeps its frequency the same as that of the DUT (Best, 2007). The manual offset should be adjusted to tune the VCO very close to the DUT frequency when the loop is not closed. Now when the loop is closed, the locking will take place with a very small average value out of the filter; i.e., the signals are in quadrature. The cutoff frequency of the low-pass filter is very important in this application. Let us suppose that it is 100 Hz. This means that the tuning voltage at the VCO varies on a timescale of ~10 ms and not faster. This imposes additional phase variation of the VCO output—this varies over a scale of ~10 ms and higher. So, the final display will have additional contributions below 100 Hz, and it can be trusted to correctly display the combined phase noise of the DUT and the VCO well above 100 Hz only (this is the value of d in Figure 6.7). In two situations we can extract information about the DUT alone:

1. If the DUT is much more noisy than the VCO, then we can identify the final spectrum observed at the phase noise of the DUT.
2. If the DUT is also a VCO and identical to the VCO used in the loop, then we can ascribe an equal contribution to the final noise spectrum from each, and the contribution of the DUT alone will be half of the values observed.

6.5.4 Analog Delay Line Discriminator Method

In this method the need for a reference source is eliminated, and the signal from the DUT is split into two channels; as shown in Figure 6.11, the signal path in one channel is delayed relative to the other. The quadrature of the two inputs is achieved by adjusting the delay line or the phase shifter, and the phase noise is obtained similar to the above method.

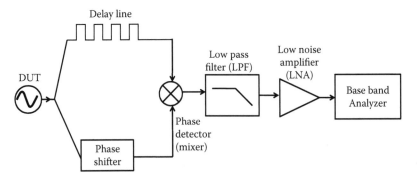

FIGURE 6.11
Block diagram of analog delay line discriminator method.

The idea here is that the phase noises at the output produced at steady state, by the DUT over two time intervals that are widely separated, are not correlated but have identical spectra. So, the spectra are added (i.e., doubled) by the mixer. Consequently, the displayed spectrum can be halved to obtain the desired phase noise. Once again, very low frequency components of the phase noise cannot be measured by this technique, since the delay required for such components to become uncorrelated will be excessive—evidently if we have a ~10 Hz noise signal and its delayed version, the delay is >>0.1 s for these to become uncorrelated.

6.5.5 Two-Channel Cross-Correlation Technique

This technique combines two phase-locked loop (PLL)-based systems (shown in Figure 6.10) and performs cross-correlation operation between the outputs of the two channels. The DUT noise is coherent, whereas the internal channel noises are canceled by cross-correlation. The setup is shown in Figure 6.12.

Here, suppose the input to the cross-correlator from the upper branch is $U(t) = K(\phi(t) - \theta(t))$, and from the lower branch it is $V(t) = K(\phi(t) - \alpha(t))$. Here, $\phi(t)$ is the phase noise of the DUT, which is identical in the upper and lower

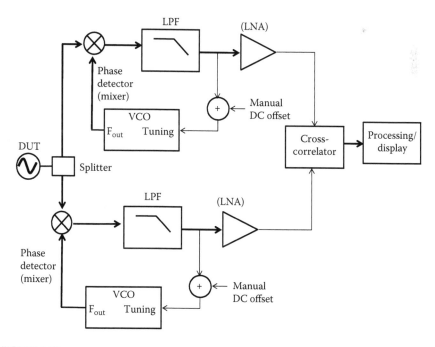

FIGURE 6.12
Two-channel cross-correlation technique.

branches, and $\theta(t)$ and $\alpha(t)$ are the phase noises of the two VCOs. Obviously, $\phi(t)$, $\theta(t)$, and $\alpha(t)$ are uncorrelated. The cross-correlator performs the function

$$R(\tau) = \int_0^T U(t)V(t-\tau)dt$$

where T is some large interval.

$$= K^2 \int_0^T (\phi(t) - \theta(t))(\phi(t-\tau) - \alpha(t-\tau))dt$$

Three of the four terms integrate to 0 since $\phi(t)$, $\theta(t)$, and $\alpha(t)$ are uncorrelated. Finally,

$$R(\tau) = K^2 \int_0^T \phi(t)\phi(t-\tau)dt$$

This is just the autocorrelation of $\phi(t)$, and its Fourier transform will show the desired PSD. The cross-correlator is actually implemented through digital signal processing (DSP), and the inputs to the cross-correlator are first digitized with an analog-to-digital convertor (ADC). Since all this operates at low frequencies (phase noise is rarely measured beyond 10 MHz offset), these systems present no challenges. This system offers the best solution to measuring phase noise, although it requires many parts and can be costly. Also, to adequately remove the unwanted terms by cross-correlation, the length of the operation has to be very large (T was assumed large above). This may make it slow.

Problems

1. The noise figure of a 20 dB amplifier is 3 dB. A defect now causes the input connector to behave as a 10 dB matched attenuator. What is the new noise figure, assuming a temperature of 290 K?

2. What is the noise figure of a 10 dB matched resistive attenuator.

3. A 10 dB gain (constant in the band considered) amplifier with a 50 Ω resistor at 580 K connected to the input gives 100 pW noise at the output, measured in a 1 GHz bandwidth. What is the noise figure? The amplifier and all instruments used are well matched to 50 Ω.

4. A 20 dB attenuator consists of a resistive tee, as shown in Chapter 3, using 41 Ω for the series resistors and 10 Ω for the shunt resistor. What are the noise figures of this when kept at 100°C and at 17°C? Recall that a resistor R can be modeled with an additional series voltage source of root mean square (rms) value $\sqrt{4kTBR}$ for noise calculations.

5. What is the output spectrum if a voltage source $A\cos(\Omega t + \phi(t))$ with a 50 Ω series resistance (this has noise) is passed through a circuit whose output is the cube of the input? The load is also 50 Ω.

References

Agilent. Fundamentals of RF and Microwave Noise Figure Measurements. Application Note 57-1. 2010.

Best, R. *Phase-Locked Loops: Design, Simulation, and Applications.* McGraw-Hill Professional, New York. 2007.

Ha, T.T. *Solid State Microwave Amplifier Design.* John Wiley & Sons, New York, 1981.

Maas, S. *Microwave Mixers.* Artech House, 1993.

7

Microwave Signal Generation

Until now we have discussed many measurement techniques, and it is obvious that all are dependent on the availability of a source with a precisely known frequency. We will now describe how such signals can be generated.

Two types of sources have been used for microwave measurements. The first (older) type is the sweeper, which is essentially a voltage-controlled oscillator (VCO) with a very large frequency range. The yttrium-iron-garnet (YIG) technology described in Chapter 2 has been used to realize such sources. The frequency of such sources is continuously variable, but this also means that the stability of the frequency is limited—for example, noise in the controlling voltage will automatically translate to random frequency fluctuation.

The second category of courses is the synthesized type, where the output is derived from a fixed-frequency (say, f_0) stable source according to the formula

$$f_{out} = f_0 \, (M/N) \tag{7.1}$$

Here M and N are integers that can be controlled through a computer. If M and N are permitted to have large values (we will see the details later), it can be seen that the frequency resolution or step size can be made very small, and for all practical purposes, such a source is no different from a sweeper. The important difference is that by virtue of (7.1), the output is frequency and phase locked to the reference frequency f_0, and hence its frequency (or phase) fluctuation is almost as low as in f_0.

As real-world examples, a C band VCO from Mini-Circuits, Inc. (part ZX95-5400+) has phase noise −58 dBc/Hz at 1 KHz offset, while the synthesized source Agilent N5171B has a resolution of 0.001 Hz and shows a phase noise of −95 dBc/Hz at 1 KHz offset when set to 6 GHz output. This huge difference in the spectral purity has led to the exclusive use of synthesized sources for measurement today. Of course, the reference frequency generator is an important component. A quartz crystal-based source is used in most equipment (often kept in a temperature-controlled environment using a heating coil and a solid-state temperature sensor—this arrangement is called the oven-controlled crystal oscillator (OCXO)). For specialized applications requiring even greater accuracy, a rubidium- or cesium-based atomic clock can be used, such as the Microsemi Timecesium 4400 (Microsemi) or precision clock signals available from navigation satellites such as GPS.

In the subsequent sections we will give further details on the concepts mentioned above.

7.1 Oscillator Circuits

Details about oscillator circuits such the Colpitts oscillator shown in Figure 7.1 can be found in any textbook (e.g., Millman and Grabel, 1987). The circuit in Figure 7.1 may appear to be different from the textbook circuits, but the differences are actually minor: a 200 pF capacitor has been added in series with the inductor to block dc and allow proper biasing, while the base-emitter capacitance of the transistor has been used to complete the conventional C-L-C resonant circuit of a Colpitts oscillator. V1 is a step voltage used for the main dc source, to simulate the oscillator turning on. Using a genernal-purpose 2N2222 n-p-n bipolar junction transistor (BJT), it was seen through SPICE simulation that oscillation could be obtained up to almost 1 GHz (practical values are, of course, expected to be much smaller, 100 MHz being a reasonable upper limit—notice that the 1 nH inductor will be difficult to realize in practice).

The interesting feature is seen in Figure 7.2, where the output voltage (at the transistor collector) is plotted. The collector voltage ultimately settles to a sinusoid with a peak-to-peak swing of about 1 V, but from the moment the supply (Vcc in Figure 7.2) is turned on, it takes around 4 μs for the output to stabilize to a sinusoidal waveform (about 4000 cycles with a 1 GHz output frequency)—even after the onset of oscillations, it takes around 2 μs to stabilize. Because of the high oscillation frequency, individual cycles are not visible in the figure if the settling time of 4 μs is to be displayed. The conclusion from this example is that if this oscillator is converted into a VCO (by replacing the 40 pF capacitor with a varactor, for example), a settling time of at least 2 μs should be allowed between frequency changes. Alternatively, if

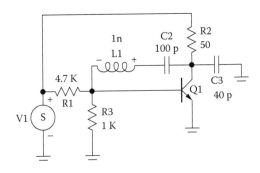

FIGURE 7.1
A Colpitts oscillator circuit.

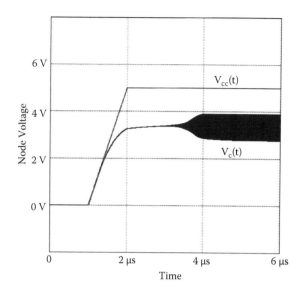

FIGURE 7.2
Simulated voltages in the circuit.

the VCO is to be tuned by an analog waveform, this waveform should not vary faster than 500 KHz.

7.2 The Crystal Oscillator

For any oscillator, the basic requirements are a gain element and a resonator element (which decides the frequency). A convenient gain element (frequently used in actual systems) is shown in Figure 7.3.

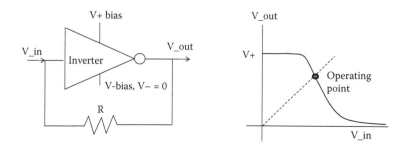

FIGURE 7.3
(a) Amplifier circuit. (b) Input-output characteristics.

The inverter is a usual digital circuit having high input impedance, and the transfer characteristics shown in Figure 7.3(b). Since the input impedance is high (several MΩ compared to R, which is ~10 KΩ), the current through R, and hence the drop across R, is negligible. This gives V_in = V_out, which is plotted as the dashed straight line in Figure 7.3(b). The intersection of this dashed line with the transfer characteristics of the inverter gives us the dc operating point. If we consider high-frequency small signal variations around this operating point, we notice a large negative slope of the transfer characteristics, which is interpreted as a high negative gain value. Notice that the dashed straight line in Figure 7.3(b) need not be followed anymore because this was arrived at considering negligible current in the resistor, which is not true if external circuitry is now connected (say, dc bypassed with capacitors) and high-frequency currents flow in R. The transfer characteristics of the inverter have to be followed as long as we do not exceed the operating frequency of the inverter.

To realize a crystal oscillator using this circuit, let us first see the equivalent circuit of a crystal. As can be seen in any textbook (e.g., Millman and Grabel, 1987), a fairly accurate equivalent circuit of a quartz crystal is given by a series L-C-R, with another C in parallel across this, as in Figure 7.4 (the circuit between ports 1 and 2).

The component values shown in Figure 7.4 are typical around 1 MHz, and give the $|S_{21}|$ shown in Figure 7.5. Notice the very high value of the inductor and the very low value of the series capacitor—as is obvious, these values have nothing to do with any actual inductance and capacitance in the physical structure of the crystal. These merely fit the observed terminal behavior.

It is seen from Figure 7.5 that there are both series resonant and parallel resonant behaviors shown by this component. The series resonant behavior is mostly used in oscillator circuits.

A crystal oscillator circuit using complementary metal oxide semiconductor (CMOS) inverters is shown in Figure 7.6. Here, two inverter-based gain

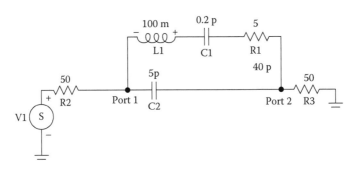

FIGURE 7.4
Equivalent circuit of a quartz crystal.

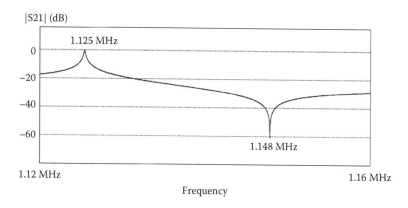

FIGURE 7.5
Frequency response of the equivalent circuit of the crystal.

blocks are cascaded to both realize higher gain and achieve a positive gain (i.e., phase shift = 0). An oscillator can also be realized using a single stage, but this circuit is more robust. The MOSFETs used for simulation (2SK10 and 2SJ10) are actually obsolete power devices, so the component values in the crystal equivalent circuit have also been changed arbitrarily to lower the operating frequency. The oscillating waveform (drain of M3) and the main supply waveform (V3 in Figure 7.6) are shown in Figure 7.7.

It is seen that the circuit indeed oscillates at 110 KHz, the series resonant frequency of the crystal model. By no means is this the only circuit for a crystal oscillator—many other circuits (for example, circuits built around a single BJT) perform equally well. The inverter-based circuit is also often

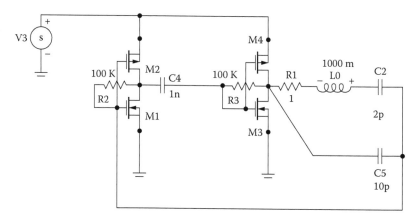

FIGURE 7.6
Crystal oscillator circuit.

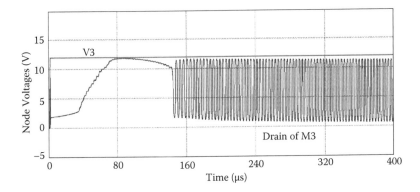

FIGURE 7.7
Simulated waveforms for crystal oscillator.

implemented using a single inverter (the Pierce-gate oscillator) using some additional components.

It is worth noting that high-quality VCOs today have fairly high spectral purity compared to simpler and less expensive types that are widely available. For example, the DCSR100-5 from Synergy Microwave Corp. USA shows a phase noise of –105 dBc/Hz at 1 KHz offset around 100 MHz output, which is similar to a synthesized source output (such as the Agilent product mentioned earlier). A high-quality 100 MHz OCXO, of course, is still superior at –150 dBc/Hz at 1 KHz offset. This is at a single frequency—a synthesized source built from this will have a degraded performance.

7.3 Tunable Oscillator

While swept frequency sources are rarely used today for measurements, a tunable oscillator is still required as an important component in some synthesized sources, as we shall see shortly. Both YIG-based oscillators and varactor-tuned oscillators are used. There is small confusion in the terminology used in this area. A varactor is a true voltage-controlled component, since an applied voltage reverse biases a diode junction, and thus changes its small signal capacitance. A YIG, on the other hand, is a current-controlled resonator, since it is the magnetic biasing using coils that change its resonant frequency. However, the term *VCO* is commonly used for either type of oscillator as long as a significant continuous variation can be obtained in the output frequency in response to an analog control signal (current or voltage).

The basic circuit of a monolithic VCO using N-MOSFETs is shown in Figure 7.8. This is the starting point for most popular MMIC oscillators (CMOS or GaAs). Varactors are used to tune the frequency as shown; the

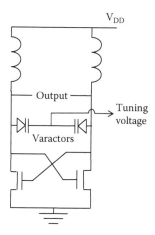

FIGURE 7.8
Simplified circuit of a monolithic VCO.

tuning voltage ensures that they are reverse biased. The difficulty with using this circuit as shown is that the resonant networks formed by the inductors, the gate capacitances, and the varactors are low Q and noisy. In many cases it may also be difficult to realize suitable-valued inductors, as these occupy a lot of area on a chip. Numerous enhancements to this basic circuit have been proposed (Razavi, 2011), and today this circuit is the core of most MMIC/RFIC (radio frequency integrated circuit) VCOs and systems using them.

A circuit (there are many types) of a YIG-tuned VCO is shown in Figure 7.9. The circuits discussed in Chapter 2 were of the transmission type; i.e., at the resonant frequency the signal passes from one port to another through the YIG. Here the resonator is used in the reflection mode; at resonance the signal is reflected by the YIG, while off-resonance it passes from one port to another as if the YIG is not there. This behavior is utilized to provide feedback to the BJT (it is used in common-base mode, although bias networks are not shown here). Values of Z1 and Z2 are optimized in the design to achieve

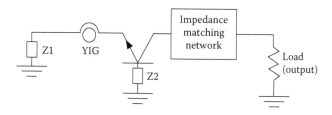

FIGURE 7.9
Simplified circuit of a YIG-tuned VCO.

the desired feedback. The impedance matching network ensures that maximum power is delivered to the load, while maintaining steady oscillation.

Some features of these circuits worth remembering are:

Monolithic VCO:

1. Small size and easily integrated with other circuits (by definition, monolithic)
2. Can operate at very high frequencies (millimeter-wave circuits in CMOS-RF are well known)
3. Very fast tuning (i.e., frequency changes quickly in response to a change in tuning voltage)
4. Low power consumption; negligible power required for tuning
5. Limited tuning range
6. Relatively high phase noise

YIG-tuned VCO:

1. Large size and weight, mainly due to tuning coils
2. Difficult to realize beyond 20 GHz
3. Slow tuning—the coils have large inductance, and it takes time for the tuning current to settle to the desired value
4. High power consumption, especially in the tuning coil
5. Very large tuning range (maximum to minimum frequency ratio easily exceeds 4:1)
6. Lower phase noise than monolithic circuits

By no means are these the only types of VCOs. Plenty of information on this subject is available in books (Rohde and Rudolph, 2013), research papers, and widely available data sheets and application notes.

7.4 Direct Digital Synthesis (DDS)

Also called numerically controlled oscillator (NCO), this is the conceptually simplest technique for synthesizing a sinusoidal waveform with certain selectable frequencies. To understand the idea behind a DDS, consider the expression

$$V(t) = \sin(2\pi f t) \tag{7.2}$$

This function of time t is actually a composite of two simpler functions:

1. $x(t) = (2\pi f)\, t$, which is a ramp function with a selectable slope.
2. $\sin(x)$, which is a nonlinear but memoryless function. Using digital techniques, such a function is easily implemented as a lookup table (commonly a read-only memory (ROM)).

It is clear that concepts from digital electronics find a natural application here. In fact, the DDS is a component that straddles the border of what were traditionally regarded as separate analog and digital techniques. Today, of course, these domains have merged.

Let us translate the above into discrete time, by assuming that the system is operated with a clock of time period Δt (which is very small) and all waveforms change at intervals of Δt. So, (7.2) becomes

$$V(n\,\Delta t) = \sin(2\pi\, f\, n\, \Delta t) \tag{7.3}$$

This is simplified to

$$V(n) = \sin(2\pi\, k\, n) \tag{7.4}$$

If we implement this using digital components (the analog voltage V will of course be represented by a number of bits), the central question becomes: What are the values that k can take? If we use a digital-to-analog converter at the end to generate the analog waveform $V(t)$, the final frequency of the analog waveform will be $(k/\Delta t)$. A block diagram implementing (7.4) is shown in Figure 7.10.

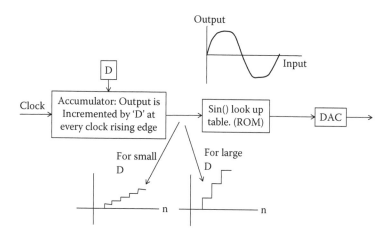

FIGURE 7.10
Implementation of rudimentary DDS.

A simple example will clarify the operation. Let us suppose that $k\,n$ in (7.4) is represented by 4 bits. Then the argument of the sin() function in (7.4) can take 16 different values.

We will have to divide the interval 0 to 2π into 16 equal parts, and store the values of $\sin(x)$ for 16 values of x equi-spaced in the interval $(0, 2\pi)$ in a ROM. So, $2\pi\,k\,n$ should take the values

$$0, (2\pi/16), 2(2\pi/16), 3(2\pi/16), 4(2\pi/16), \ldots, 15(2\pi/16)$$

Now, remember that $k\,n$ must represent a ramp function. So, the simplest possibility is that $k\,n$ takes the values

$$0, (2\pi/16), 2(2\pi/16), 3(2\pi/16), 4(2\pi/16), \ldots, 15(2\pi/16), 16(2\pi/16), 17(2\pi/16), \ldots$$

Since the sin() function is periodic with period 2π, the ramp function can always make jumps of 2π, and the sequence above becomes

$$0, (2\pi/16), 2(2\pi/16), 3(2\pi/16), 4(2\pi/16), \ldots, 15(2\pi/16), 0(2\pi/16), 1(2\pi/16), \ldots$$

This is known as the sawtooth function. Regardless, $k = 1/16$ in this case.

In actual implementation, the 4-bit number representing $k\,n$ will be incremented with a particular constant value at each count of the clock, and will always wrap around with an overflow after the bits reach 1111.

Another possibility is the sequence

$$0, 2(2\pi/16), 4(2\pi/16), \ldots, 14(2\pi/16), 0(2\pi/16), 2(2\pi/16), \ldots$$

Now $k = 2/16$.

A complication results when we try to set $k = 3/16$. Now the sequence becomes

$$0, 3(2\pi/16), 6(2\pi/16), 9(2\pi/16), 12(2\pi/16), 15(2\pi/16), 2(2\pi/16), 5(2\pi/16), \ldots.$$

While the waveform for $k = 2/16$ is obviously a frequency-doubled version of the waveform for $k = 1/16$, it is certainly not correct to extend the reasoning and say that $k = 3/16$ gives a frequency-tripled version of the waveform with $k = 1/16$.

Let us plot these waveforms—this is shown in Figure 7.11.

It is obvious that the waveform starts deviating from the desired sinusoidal shape at $k = 3/16$, and is completely unusable for $k > 7/16$. One important conclusion is that the maximum frequency should be kept well below $(0.5/\Delta t)$. Also, the frequency resolution is $[2^N \Delta t]^{-1}$, where N bits are used, if the simple scheme outlined above is used. In reality, a modified version is used to get good resolution without using an excessive number of bits for the ROM address—a ROM with many address bits is slow and would severely limit the clock, and thus the upper frequency limit.

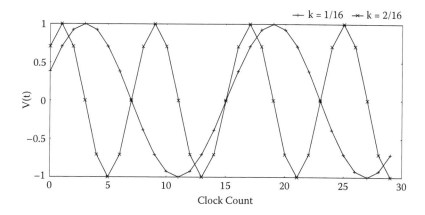

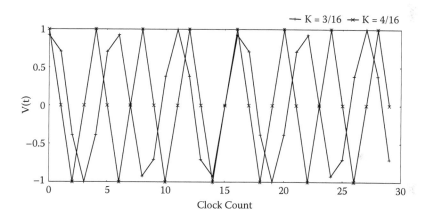

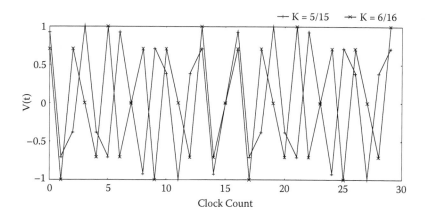

FIGURE 7.11
Waveforms for different *k*. (*continued*)

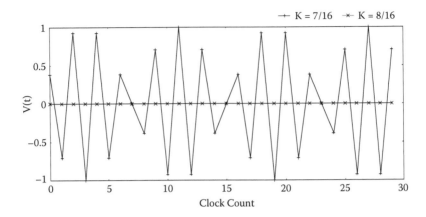

FIGURE 7.11
(*continued*) Waveforms for different k.

This brings us to the basic architecture of a DDS, shown in Figure 7.12. The register transfers its input to output at every clock rising edge, and then holds it fixed until the next rising edge.

Looking at Figure 7.12 we notice:

1. The number of bits should be high to achieve good frequency resolution.
2. Not all the bits should be used as ROM address bits—the highest few bits most significant bit (MSB) will suffice. Note that one time

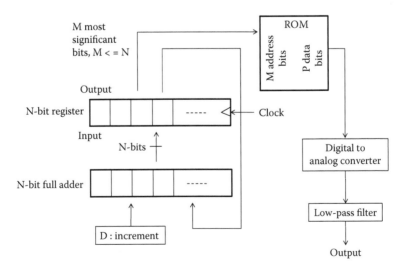

FIGURE 7.12
Improved architecture of a DDS.

period of the output is equal to the time taken between adder overflows. This is how the register gets reset (say, to all zeros). Automatically, this is also the time between MSB resets (or resets of the most significant few bits).

3. Up to the DAC, everything is digital; so, the values of the sin() are digital representations, not actual values. The actual analog waveform is generated from the digital to analog convertor (DAC). Even that is a staircased approximation. If required, the low-pass filter can smooth this to a close approximation to a true sin() curve. The number of bits P used to define the sin() function is not really important—typically it ranges from 8 to 12.

4. The frequency step size is $[f_{clock}/2^N]$ and the maximum frequency is usually kept below $f_{clock}/3$. The frequency output for some D is $D[f_{clock}/2^N]$.

5. The output signal is ultimately derived from the clock. So, it is imperative that a highly stable low-phase noise clock is employed. This is usually obtained from a crystal oscillator using a multiplier chain, or a fixed phase-locked loop (PLL) with a high dividing factor (prescaler), as we will see later.

As an example, Figure 7.13 shows the spectrum of the output waveform for $N = 16$, $M = 8$, $D = 9555$, and clock frequency = 1 GHz. The "spurs" are caused by distortions in the waveform as in Figure 7.11(b, c). Here, of course, the distortion is much smaller because the number of bits is larger. However, the number of bits being finite, there is a rounding-off or truncation error when the bits are reduced from N to M. Moreover, this error has a certain periodicity—Figure 7.11 shows an exaggerated case of this. This gives rise to the spurs.

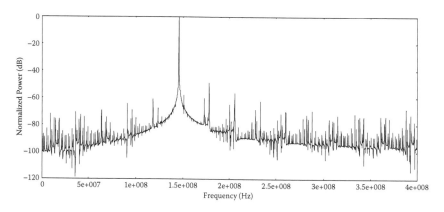

FIGURE 7.13
Spectrum of DDS output.

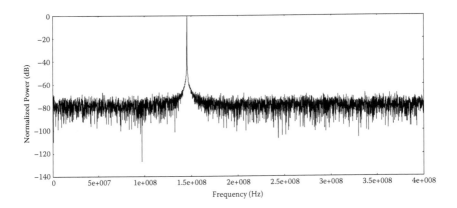

FIGURE 7.14
Spectrum of DDS output with randomization.

This can be further improved by a simple trick. The 16-bit output of the register can be incremented by a *random* number lying between –0000000011111111 and 0000000011111111. The most significant 8 bits are taken after that. There is very little visible change in the waveform if this is done, but the spectrum is cleaned up dramatically, as shown in Figure 7.14. Essentially, this takes the power in the spurs and spreads it over a wide spectrum, thereby eliminating the spurs.

The biggest advantage of the DDS is that change in output frequency is practically instantaneous—if D is changed, the output frequency changes from the very next clock cycle. Also, the change is smooth—there is no abrupt change of phase. This speed in turn stems from the feed-forward architecture—note that there is no feedback loop.

Incorporating some additional elements into the basic architecture of Figure 7.12, many additional functions may be implemented. For example, a message signal in digital form may be used for D—this will directly give an frequency shift keying (FSK) generator. Alternatively, the adder may incorporate an additional input, which is normally 0 but may become nonzero for one particular clock cycle. This will shift the phase of the output in accordance with what that nonzero value was. A PSK modulator results from this.

To conclude, the DDS is particularly useful in military applications, where very fast frequency switching is required. The major limitation is that it cannot operate beyond a few GHz, as digital systems that are clocked beyond this speed are not easy to realize. Also, the output is not truly analog (although it is filtered—the filter will definitely remove the prominent clock frequency from the output), so if spectral purity is paramount, a DDS may not suffice. There are some additional issues related to output filtering—these are advanced topics that can be found in references like Rohde (1997).

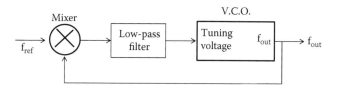

FIGURE 7.15
Basic PLL.

7.5 PLL-Based Synthesizers

These are the preferred microwave sources for instrumentation, when switching speed is not critical. The heart of this is the simple phase-locked loop (PLL) shown in Figure 7.15.

Before going on to describe the operation of this, it will be desirable to focus on the mixer in Figure 7.15. A traditional analog (diode-based) mixer can certainly be used here, but another class of mixers that is frequently used in such applications is based on the XOR/XNOR (exclusive OR/NOR) gates. For example, Figure 7.16 shows the result of applying two binary waveforms of different frequencies to an XNOR gate. To bring out similarity with traditional multiplication, +1 and –1 levels are shown instead of the usual 1 and 0 levels. Obviously, level shifting at the physical level can be easily done with resistors and a battery.

It is obvious from Figure 7.16 (although the frequency difference here is exaggerated) that effective multiplication is achieved, and the low-pass filter will pass the average value shown—a slowly fluctuating signal with frequency = difference of the input frequencies. One interesting feature here is that local oscillator (LO) and RF are equally strong—unlike traditional analog mixers where RF strength is small. This does not cause any difficulty with harmonic mixing for this digital mixer.

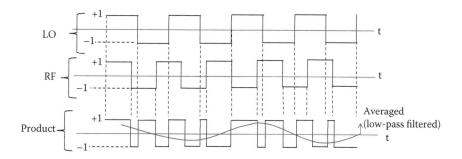

FIGURE 7.16
Mixing using XNOR gate.

The operation of this PLL is well known: assuming that the VCO free-running frequency is not very different from the input frequency f_{ref}, we will get a slowly varying signal at the control input of the VCO. This will cause the output frequency to fluctuate. While fluctuating, if the output frequency becomes equal to f_{ref}, the difference frequency becomes 0. This means that the tuning voltage of the VCO will stop fluctuating and stabilize at a suitable dc value, which will lock the output to f_{ref}. At this stage, let us use the following notation:

$$V_{ref} = A\cos(\omega_{ref}t), \quad and \quad V_{out} = A\cos(\omega_{out}t + \theta) \tag{7.5}$$

Then ω_{ref} being equal to ω_{out}, we have $V_{tune} = K\cos(\theta)$.

Any subsequent fluctuation in θ will cause a change in the tuning voltage that will counter this phase fluctuation. The output is hence locked to the input. This is not obvious, but phase-locked loop analysis is well known (e.g., Rohde, 1997), and such analysis confirms that indeed this locking will happen; moreover, if there is a change in f_{ref}, this analysis also reveals that the output will lock to this new reference after some delay governed by the VCO, mixer, and low-pass-filter properties. The details are beyond the scope of this book.

Now it appears that we have not achieved anything—the circuit has just replicated the input at the output. The real strength of the circuit is apparent once we add a few simple components to the loop. Consider the modified circuit shown in Figure 7.17.

A programmable divider in its simplest form is a digital counter with a few additional gates that reset the counter whenever the count reaches $N-1$, referring to Figure 7.17. The clock of this counter is f_{out}, and this reset pulse is the output of the divider that is fed back to the mixer. Of course, this is not a very good divider because the output will be a train of pulses with a low duty cycle of 1/16 when $N = 16$, but it demonstrates the concept. We have thus shown a frequency synthesizer that can generate frequencies 10, 20, 30, …,

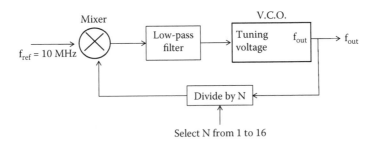

FIGURE 7.17
Rudimentary PLL-based synthesizer.

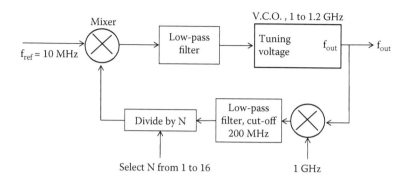

FIGURE 7.18
Synthesizer with frequency offset.

160 MHz. These are phase locked to the reference frequency, and hence they have the same phase noise (roughly—check Rohde [1997] for more information). Of course, we have assumed that the VCO can generate signals from 10 to 160 MHz, which is not practical.

Another modification is shown in Figure 7.18. Now we see that $(f_{out} - 1$ GHz)*N is locked to 10 MHz. This tells us that frequencies that can be generated are 1010, 1020, ..., 1160 MHZ. This range is reasonable for a VCO.

We can further subtract the same 1 GHz with a mixer from the output, giving us, again, the 10 to 160 MHz range.

Let us now see how the above concepts are used to synthesize frequencies from 2 to 8 GHz with a resolution of 1 MHz, in a practical synthesizer. The particular instrument considered is the microwave synthesizer GT9000 from Gigatronics, Inc. This product uses 3 YIG VCOs to cover the range 2 to 26 GHz, an additional YIG to generate some discrete frequencies to use internally in the synthesis process, and some additional circuits to generate frequencies below 2 GHz. Let us see the operation for the 2–8 GHz range.

The circuit that generates the output signal for this range is shown in Figure 7.19. The YIG VCO has a range of 2–8 GHz. This is divided into bands, and for each band a preselected f_{ref} is stored in the computer memory. Some of these bands (in MHz) are shown in Table 7.1.

Generation of these reference frequencies requires a separate subsystem that will be discussed later—for the moment let us assume that any one of these can be generated in response to a command from the computer.

The starting point is a crystal oscillator generating 10 MHz. This is divided by a fixed value of 10 to obtain a 1 MHz reference (seen at the left in Figure 7.19). The PLL technique ensures that the output of the "divide by N" element results in 1 MHz. N can be set to values 5, 6, 7, ..., 394, 395 by the computer. This means that the output frequency is above f_{ref} by a value that can be set to 5, 6, ..., 395 MHz. As a specific example, let us suppose that

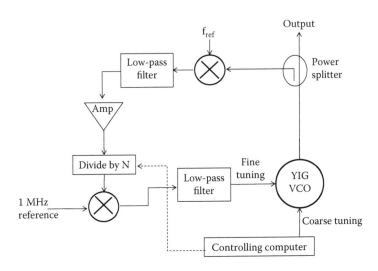

FIGURE 7.19
Output generating portion of the synthesizer, using a 2–8 GHz YIG VCO.

an output of 2192 MHz is required. The computer accordingly does three things:

1. It sets f_{ref} to exactly 2005 MHz.
2. It sets the coarse-tuning current to tune the YIG to 2192 MHz. Of course, this is only coarse tuning, so at this stage the output frequency may be 2194.
3. It sets $N = 187$.

Assuming that $f_{out} = 2194$ MHz, we get a divide by N output at (2194 – 2005)/187, or 1.0107 MHz, which is not quite 1 MHz. So, a 0.0107 MHz signal will go to the fine-tuning control and trim the output frequency until the output settles at exactly 2192 MHz, and the fine-tuning current becomes steady. Note that the coarse tuning ensures that the VCO does not settle at 187 MHz below f_{ref}.

TABLE 7.1

Reference Frequencies for a Few Bands

Band	f_{ref}
2001–2099	1905
2100–2199	2005
5200–5299	5105
7900–7999	7805

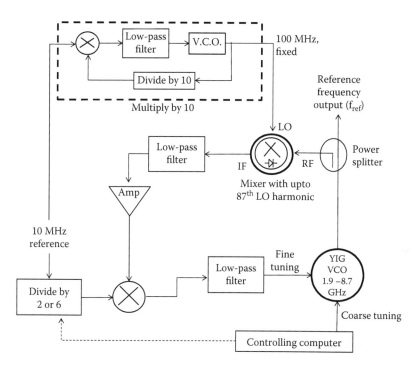

FIGURE 7.20
Generation of the reference frequencies.

For this band (2–8 GHz) values of N beyond 194 are not used, but in other bands higher values are required. For example, to generate 8199 GHz using a different YIG, an f_{ref} of 2601.667 MHz is required. The third harmonic of this mixes with the output, and using $N = 394$, proper lock is obtained.

The circuit for generating the suitable reference frequency (f_{ref} in Figure 7.19) is shown in Figure 7.20.

The key element here is a harmonic mixer (analog) that uses a step recovery diode to generate LO harmonics up to the 87th. The LO is fixed at 100 MHz. The intermediate frequency (IF) is locked to 10/2 = 5 MHz or 10/6 = 1.667 MHz. Suppose the computer has to generate 2105 MHz. It then sets the coarse tuning to this frequency; this will bring it close, say, to 2104 MHz. This will mix with the 21st harmonic of 100 MHz to yield 4 MHz. The 10 MHz reference is divided by 2 in this case to yield 5 MHz, and a difference of 1 MHz will reach the fine-tuning control of the YIG, forcing the output to rise to 2105 GHz, at which point lock will be achieved.

The important feature of the whole scheme is that everything is derived from the single 10 MHz reference—so the final signal will have a phase noise of the same order as this reference, as in all synthesized sources. This approach of first generating a set of frequencies in one loop, and then using these frequencies in another loop to achieve high-resolution in bands, is called dual-loop synthesis; obviously, more loops can be employed to obtain better range or resolution.

This particular product is somewhat outdated today; modern synthesizers usually have a <1 Hz frequency resolution. One of the techniques used to achieve this is fractional-N synthesis.

7.6 Fractional-N Synthesis

Going back to Figure 7.17, suppose that the number N is not fixed, but varies between, say, 10 and 11. To be more precise, the counter gives a pulse output after 10 input clock cycles (from f_{out}); then the next pulse comes after 11 clock cycles, then again after 10 clock cycles, and so on. In this case, the output will be locked to 105 MHz. This can be justified in the following way.

Suppose the output is indeed 105 MHz (time period 0.00952 μs). Then the output pulses from the divider are spaced by alternately 10*0.00952 μs and 11*0.00952 μs. The fundamental period of such a periodic waveform is 21*0.00952 μs; i.e., its frequency is 105/21, or 5 MHz. The second harmonic of this will mix with the reference frequency and produce a dc that will indeed lock the VCO. Proceeding along the same lines, dividing by 11 to obtain nine pulses at the divider output and then dividing by 10 to obtain the next pulse, and continuing this, will provide an effective division ratio of (11*0.9 + 10*0.1), or 10.9. However, this technique requires certain modifications before it can be implemented in practice (and it is indeed widely used today)—these details can be found in Rohde (1997).

7.7 Useful Components

Going back to Figure 7.17, we see that there are two components other than the VCO. One is the mixer + filter combination, and the other is the divider. Currently, the mixer + filter combination is usually replaced by a phase-frequency detector with a charge pump. The divider often takes the form of a dual-modulus prescaler instead of a simple counter. Let us see what these components achieve.

7.7.1 Phase-Frequency Detector (PFD)

It was mentioned in connection with Figure 7.15 that the tuning voltage of the VCO fluctuates when the two frequencies f_{ref} and f_{out} are not equal. This uncertain fluctuation can be improved to a well-behaved signal by using a

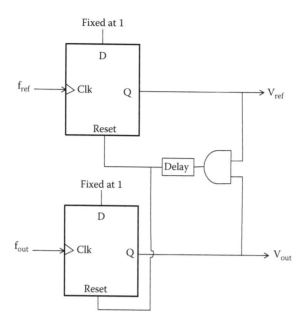

FIGURE 7.21
A phase-frequency detector.

phase-frequency detector, the most common version of which is shown in Figure 7.21. This is implemented with two D flip-flops that are rising edge triggered; i.e., the D-value (which is fixed at 1) is transferred to Q at a clock rising edge. Subsequently, Q remains fixed (at 1) until reset. This reset happens the moment (actually with a small delay, as shown) both outputs are high.

The outputs of the phase-frequency detector when two different frequencies are input are shown in Figure 7.22. The way this works is at each rising

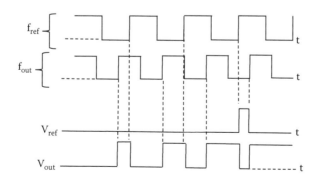

FIGURE 7.22
Typical inputs and outputs of the PFD for $f_{out} > f_{ref}$.

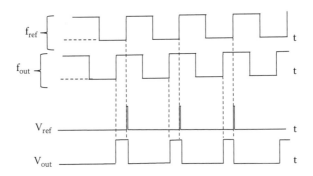

FIGURE 7.23
Inputs and outputs of the PFD for $f_{out} = f_{ref}$ but f_{out} leading.

edge of f_{ref}, V_{ref} goes high, and stays high if V_{out} is low. If V_{out} was high at that edge, then both V_{out} and V_{ref} are rest to 0 after a very small delay. A similar behavior is observed with f_{out}. So, the circuit (loosely speaking) senses whether f_{ref} or f_{out} has rising edges coming earlier and more frequently. An example is shown in Figure 7.22. It is noticed that V_{out} stays high much more than V_{ref}, corresponding to the higher frequency f_{out}.

Another case is seen in Figure 7.23. Here, the two frequencies are equal, but f_{out} leads f_{ref}. Once again, we see that V_{out} is high for more time than V_{ref}—the narrow pulse in V_{ref} actually corresponds to the small delay incorporated after the AND gate.

Now, if we could average the difference of V_{ref} and V_{out} and feed this to the VCO tuning input, it would appear that this would force the VCO to adjust its output until f_{ref} and f_{out} are locked in phase, at the same frequency. Notice that this approach will work even if the two frequencies are quite far apart—we will not end up with a rapidly fluctuating voltage at the VCO tuning input. This averaging is accomplished with a charge pump. The circuit of this is shown in Figure 7.24. In the charge pump, the important MOSFET switches are the ones connected to V_{out} (M1: an NMOS) and the one connected to V_{ref} inverted (M2: a PMOS). The rest of the circuitry implements two current mirrors that ensure that a fixed current I_0 flows through M1 or M2 if that device is ON. So, if M1 is ON, the output capacitor will discharge with a constant I_0, resulting in a steady decrease in output voltage. If M2 is ON (this will happen if V_{ref} is high), the capacitor will charge due to a constant I_0, resulting in a rise in output voltage. Both FETs are ON only for a very small duration (see Figure 7.23)—this need not be considered. When both FETs are OFF, the output voltage naturally remains steady. For the waveforms in Figure 7.23, the charge pump output is shown in Figure 7.25. Assuming that the VCO frequency decreases when the tuning voltage is raised, this charge pump output can be directly applied to the VCO, and it will force the frequency to come down slightly, but the moment this happens, the charge pump output will start going down (now that f_{out} has fallen below f_{ref}), bringing the output

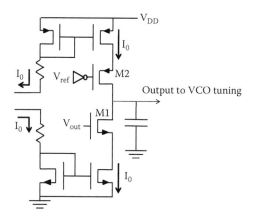

FIGURE 7.24
Simplified circuit of a charge pump.

frequency back to f_{ref}. But now the phase lead of f_{out} will be gone. Once phase match and frequency match are both achieved, both V_{ref} and V_{out} will display very narrow spikes (since rising edges always come together now), and the charge pump output will stay constant.

The above explanation is somewhat oversimplified, and issues like what exactly happens when both inputs are almost matched are discussed in books such as Best (2007) and research papers—this subject is still seeing innovations. Also, this component is simple enough to be simulated at the circuit level by commonly available simulators—this will bring out the exact behavior.

7.7.2 Dual-Modulus Prescaler

Once again, consider Figure 7.17. The divider shown is a programmable type; i.e., the division ratio can be set by the user. Such a component is relatively slow, while a fixed ratio divider (say, a divide by 10 circuit) can be operated at >1 GHz clock speeds. So, we can design a circuit like the one shown in Figure 7.26 to generate high-output frequencies.

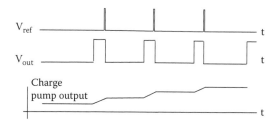

FIGURE 7.25
Charge pump output for waveforms in Figure 7.23.

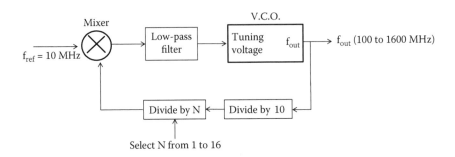

FIGURE 7.26
A prescaler (divide by 10).

The programmable divider works at a low frequency (10 to 160 MHz), while the input to the fixed divider (prescaler) goes up to 1600 MHz. Unfortunately, the step size now is 100 MHz, and we have lost resolution—basically, the overall division ratio can change in steps of 10. There is a partial solution to this, whereby two fixed-frequency dividers can be used to achieve a variable division ratio at high speeds. Such a circuit is shown in Figure 7.27. There can, of course, be many variations in the actual implementation.

In this circuit two high-speed fixed ratio dividers are used: one with divide ration M and the other with $M + 1$ (actually these numbers need not be separated by 1, but this is the prevalent configuration that gives the smallest step size). The rest of the circuit operates at low speeds. Which of these is selected is decided in the following way:

1. Initially the counters are loaded with N and A (which are programmed by the user). So, the $M + 1$ path is followed since the A-counter is not 0.

2. After $(M + 1)A$ input cycles, the A-counter would have reduced to 0 (it counts down by 1 every $M + 1$ input clock cycles).

3. At this point the N-counter would have reached a count of $N - A$ (so, A should be strictly less than N).

4. After $M(N - A)$ input clock pulses the N-counter reaches 0 and an output pulse is generated. This reloads the counters and starts the cycle afresh.

So, the time interval between output pulses is $[(M + 1)A + M(N - A)] T_{in} = [MN + A] T_{in}$. This gives the division ratio of the overall system as $[MN + A]$. Since A can be programmed to 1, 2, 3, etc., the division ratio can now vary in steps of 1, even though the high-speed components are not programmable. This circuit is widely used in frequency synthesizers.

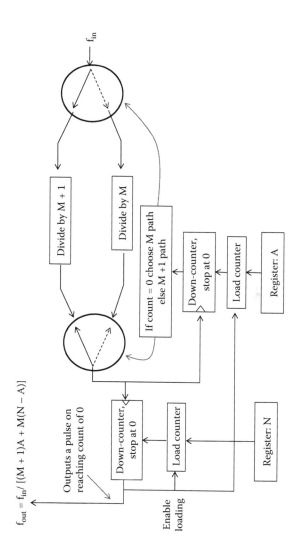

FIGURE 7.27
A dual-modulus prescaler.

7.8 Conclusion

The most commonly used techniques in microwave signal generation were described briefly. The field of frequency synthesis is still growing today, and the information provided here is only intended to give a background from which the reader can start exploring the specialized literature in this field.

Problems

1. If a signal $\cos(\omega t + \phi(t))$, where $\phi(t)$ is the slowly varying phase noise, is passed through a square-law device to generate the second harmonic, what will be the phase noise of the second harmonic?

2. If the above signal is split in two, and supplied as LO and RF to an ideal multiplier + low-pass filter, with a large delay in the RF branch, what will be the power spectral density of the output?

3. A rudimentary DDS uses a 4-bit accumulator and 4 ROM address bits. What will be the frequencies that can be obtained if there is additional circuitry that resets the counter whenever an overflow takes place?

4. Generate the waveforms when a 105 MHz pulse train is divided alternately by 10 and 11, and the resulting waveform is mixed using XNOR with a 10 MHz reference. Verify that there is indeed a dc component in the result.

5. Extend the system in Figure 7.18 to achieve synthesis of the range 10 MHz to 1 GHz in 10 MHz steps.

6. What division ratios can be obtained with a 10/11 prescaler?

References

Best, R. *Phase Locked Loops: Design, Simulation, and Applications.* McGraw-Hill, New York, 2007.

Microsemi. http://www.microsemi.com/products/timing-synchronization-systems/time-frequency-references/telecom-primary-reference-sources/timecesium-4400.

Millman, J., and A. Grabel. *Microelectronics.* McGraw-Hill, New York, 1987.

Razavi, B. *RF Microelectronics.* Prentice-Hall, New Jersey, 2011.

Rohde, U.L. *Microwave and Wireless Synthesizers: Theory and Design.* Wiley Interscience, 1997.

Rohde, U.L., and M. Rudolph. *RF/Microwave Circuit Design for Wireless Applications,* John Wiley & Sons, New York, 2013.

8

Microwave Oscilloscopes

The very moment we hear *oscilloscope* it brings a sense of familiarity to many of us. Probably, it is that instrument with which many of us have had a chance to turn the knobs, other than the multimeter in college laboratories. On the other hand, the sense of belonging soon fades away, with the adjective of *microwave* attached. Does it operate at microwave frequency? Can one really see a microwave signal at tens of GHz on the display? How does it work? This chapter is dedicated to finding answers to these questions, with emphasis on the fundamental operating principle behind *microwave oscilloscope.*

8.1 Introduction

An oscilloscope is most often used to measure and display an electrical signal (typically voltage signal at a node) as a function of time. A popular display has the voltage signal on the Y-axis against time on the X-axis, as shown in Figure 8.1. Hence, it is a time domain measuring equipment. Typically, an oscilloscope will capture the signal of interest (data) in the time domain, process the data, and display it.

Broadly, oscilloscopes can be categorized into analog oscilloscope and digital oscilloscope based on the type of processing. Analog oscilloscopes use analog processing, right from probing (or capturing the signal) to display. The processing is done in analog domain. Cathode-ray oscilloscopes, still present in many laboratories, are analog oscilloscopes. But the tremendous advancement in integrated circuit (IC) technology, with ever-increasing operating speeds, is steadily making analog oscilloscopes almost obsolete. On the other hand, digital oscilloscopes are dominating today because of their superior performance and flexibility (not to mention colorful display). Using digital oscilloscopes, one can not only display the signal of interest, but also store it and perform a variety of mathematical operations. The most exciting feature is that the user can access the signal in data form, for further detailed analysis. Hence, keeping in touch with recent technological developments, in this chapter we will be covering operating principles of digital oscilloscopes used for microwave signals.

The basic architecture of digital oscilloscopes would include an amplifier/ attenuator followed by an analog-to-digital convertor (ADC) leading to a

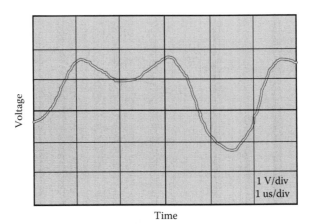

FIGURE 8.1
Typical oscilloscope display.

digital processor and finally display unit. Figure 8.2 shows the block-level description. One can also access the captured signal in data form (e.g., USB is very common).

Once the incoming analog signal is digitized by ADC, the rest of the processing is completely in the digital domain using powerful microprocessors. As a result, various math operations like peak-to-peak, root mean square (rms), maximum, minimum, mean, period, duty cycle, etc., are all performed quite easily in the digital domain. Various types of digital oscilloscopes found commercially, like digital storage oscilloscope (DSO), digital phosphor oscilloscope (DPO), mixed domain oscilloscope (MDO), and mixed signal oscilloscope (MSO), all have similar architecture. They only vary in the processing unit (Tektronix, 2014; Agilent Technologies, 2013a; Poole; Rhode & Schwarz). For example, DPO uses parallel processing for faster display than serial processing in DSO. MDO has a feature to display the spectrum of a signal as well, which is computed in the digital domain through fast Fourier transform (FFT) or its variation. In addition, MSO combines the logic analyzer embedded in the processor, which becomes handy to probe and debug

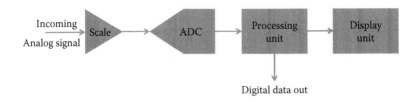

FIGURE 8.2
Digital oscilloscope (DO): typical architecture.

signals on a field-programmable gate array (FPGA) or application-specific integrated circuit (ASIC). However, the signal capture through the amplifier/attenuator, followed by sampling and digitization using ADC, is the same and common in all of them, as shown in Figure 8.2.

8.1.1 Requirement for Microwave Oscilloscope

Now coming to the microwave oscilloscope, based on previous discussion, a couple of things become clear. The ADC should be of very high sampling rate to capture the microwave signal. Also, the storage capacity or memory depth is also expected to be pretty high in order to capture sufficient signal duration. Naturally, all this adds to sophistication and expense. So, the very first question that arises is: Why would one need such a sophisticated oscilloscope, particularly at microwave frequency, where the vector network analyzer (VNA) and spectrum analyzer (SA) are already well matured?

Let us start with the spectrum analyzer. The spectrum analyzer measures the spectrum (spectral content) of the *signal* being probed. That is, it gives the power distribution across the frequency range of the signal being probed. So, it gives scalar information of the signal. The phase information is lost in the measurement process. So, in conclusion, the spectrum analyzer cannot be used to either obtain complete information of the signal being probed or characterize the system or device under test (DUT).

On the other hand, the VNA is much more powerful. It measures *S*-parameters as a function of frequency of the DUT in steady state. Hence, VNA can characterize the system or DUT in the steady state. *S*-Parameters measured are complex quantities; i.e., they have magnitude and phase. But, on the other hand, VNA cannot be used to characterize the signal. One cannot use VNA to measure either the incoming signal to DUT or the outgoing signal from DUT. Thus, VNA is more useful in system design and characterization than in debugging or signal property measurement. Even in system characterization, if we are not able to examine the input signal and the output signal independently, many aspects of verification of the designed system (e.g., a frequency doubler) for desired performance are very difficult to perform. For example, an oscillator has only an output, so a vector network analyzer is of no use (even a spectrum analyzer does not show start-up time for an oscillator).

So, VNA alone is not sufficient to characterize a microwave system, specially nonlinear or noise-related aspects. Note that VNA can characterize the DUT only in steady state and is useful only when the DUT is in the linear region of operation (since *S*-parameters are defined for linear time-invariant systems). For example, VNA cannot be used to determine the switching behavior of an ASK (amplitude shift keying) modulator at the switching instant (even though the modulator has been characterized separately for steady-state ON and OFF modes of the modulator).

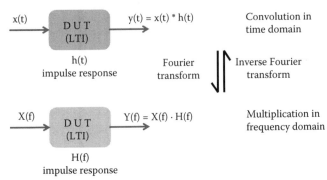

Spectrum Analyzer: **|X(f)|** and **|Y(f)|** : Frequency Domain Measurement
Vector Network Analyzer : **H(f)**: Frequency Domain Measurement
Microwave Oscilloscope: **x(t)** and **y(t)**: Time Domain Measurement

FIGURE 8.3
Microwave measurements for LTI system.

So, neither the spectrum analyzer nor VNA is sufficient to design and verify many microwave systems. This void has been felt heavily in recent times, especially when microwave systems are becoming progressively sophisticated with complex modulation schemes for ever-increasing data rates. Characterizing the signal, whether at the input to a system or at the output of a system, has become more of a necessity. Switching behavior of a DUT is no longer negligible, especially with rapidly increasing data rates. So, this is where microwave oscilloscopes fill the void and come to the rescue to realize high-performance microwave systems. Microwave oscilloscope, in addition to capturing the signal in the time domain, can also be used to obtain the transient response of switching systems for a specified input. Figure 8.3 explains the utility of each measuring instrument discussed above, i.e., spectrum analyzer, VNA, and microwave oscilloscope, in analysis, design, and verification of linear time-invariant (LTI) systems at microwave frequencies.

Table 8.1 gives a summary of the functionality and utility of the microwave oscilloscope compared to the spectrum analyzer and VNA discussed so far. We shall keep adding features to this comparison table as and when we keep exploring more.

8.1.2 Block Diagram

Now that we appreciate the utility of the microwave oscilloscope, the stage is set to explore more, to know how such an oscilloscope works. Figure 8.4 shows a simplified block diagram of a typical high-speed digital oscilloscope.

TABLE 8.1

Comparison Table among Microwave Oscilloscope, VNA, and Spectrum Analyzer

Features	Spectrum Analyzer	Vector Network Analyzer (VNA)	Microwave Oscilloscope
Signal characterization	Partial (frequency spectrum)	Not possible	Complete (time domain)
System characterization (LTI systems)	Very limited	Possible (frequency domain)	Possible (time domain)
Nonlinear system characterization (NL-TI systems)	Limited	Not possible	Best option, among all instruments

The basic functional blocks of any high-speed digital oscilloscope are:

1. **Vertical system:** A vertical system comprises an attenuator and amplifier to perform scaling operation on the analog/microwave signals to be probed. Hence, this block can also be referred to as analog front-end block. The vertical system enables the user to manually scale and position the incoming signal on the display screen as per requirement by adjusting the front panel controls like "scale," "volt per division," etc. The termination seen by the analog signal can be either 50 Ω or high impedance (typically 1 MΩ, like in low-frequency DSO). However, generally termination is 50 Ω for very high speed oscilloscopes operating in the microwave regime. Broadband behavior is the key requirement of the components of a vertical system.

2. **Horizontal system:** A horizontal system performs the functionalities of signal acquisition and conversion. It is directly related to the front panel controls, like "sampling rate" and "time per division." Internally, depending on the user front panel inputs, suitable clock signaling is generated and fed to the ADC. Depending on the

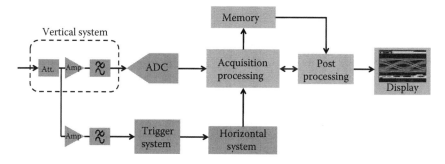

FIGURE 8.4
Block diagram of typical high-speed digital oscilloscope.

memory depth of the oscilloscope, the acquisition time (duration of signal captured or stored) is determined.

3. **Trigger system**: This is one of the key subsystems for effective utilization of an oscilloscope. This subsystem determines when to acquire the incoming signal; i.e., it initiates the oscilloscope to make an acquisition. It is an extremely vital component, particularly for achieving a stable view of repeating waveforms. The trigger signal can be derived from an incoming analog signal through well-defined conditions like positive edge, negative edge, crossing a threshold, etc. However, the user has the option to provide an external signal to act as a trigger. This feature is very useful particularly when the incoming signal is noisy.

4. **Display system**: The display system determines the signal representation to the user. It enables the user to perform various display functions, like selective zoom, overlapping, color indexing, marker options, and so on. Nowadays, most display systems in oscilloscopes also encompass the "touch screen" feature for users to interact with the controls directly on the screen.

The fundamental operation in a digital oscilloscope is signal sampling and acquisition using high-speed ADC, which is a vital component. However, the sampling rate of ADC required in order to capture a microwave signal is extremely high, extending to tens of giga-samples per second (GSps). Thus, in the following section, let us discuss more on sampling techniques used in such high-speed digital oscilloscopes (microwave oscilloscopes). Microwave oscilloscopes can be categorized broadly into two types, based on the type of sampling (data acquisition) as:

1. Real-time oscilloscope
2. Sampling oscilloscope

8.2 Overview of Real-Time Oscilloscope and Sampling Oscilloscope: Real-Time Oscilloscope (Single-Shot Oscilloscope)

The working principle of the real-time oscilloscope is straightforward to grasp. Basically, a very high speed ADC is used to capture the incoming signal once the trigger is applied. Trigger is a kind of signal that tells the oscilloscope when to digitize and store the data in memory. A trigger can be

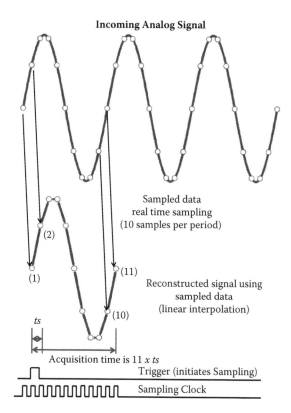

FIGURE 8.5
Real-time oscilloscope: operating principle.

derived from the incoming signal itself through precise definition, like positive edge, negative edge, attaining a certain threshold value, etc. An external signal can also be fed to act as a trigger. Once a trigger occurs, ADC starts sampling, digitizing, and storing the sampled data in memory, for further processing and display. Figure 8.5 gives a clear pictorial representation.

In Figure 8.5, when a trigger occurs, the ADC samples and stores the incoming signal continuously in memory, until the next trigger occurs. As per the Nyquist sampling theory the ADC sampling speed should be at least twice the highest frequency of interest. The memory depth determines the acquisition time interval.

Example 8.1

A commercial real-time digital oscilloscope has a sampling rate of 2.5 GSps. Thus, any incoming signal with bandwidth greater than 1.25 GHz would cause aliasing as per the Nyquist sampling theory.

Example 8.2

A commercial real-time digital oscilloscope has a sampling rate of 2.5 GSps and memory depth of 250 Mpts (mega-points). Thus, the maximum time interval that can be captured or acquired, operating at 2.5 GSps, is 0.1 s, i.e., (250 M/2.5 G). If each point is stored as 1 byte, then the memory required is 250 MB.

Now let us see how this operation translates in the frequency domain. Figure 8.6 gives an illustration. For the sake of simplicity, impulse sampling is assumed. Actually, there will be distortion due to finite sampling pulse and quantization, which we will ignore for now, illustrating the principle of operation. The incoming signal is denoted as $x(t)$ in the time domain, and let it have the highest frequency of interest as f_m Hz. The impulse sampling signal is denoted as $y(t)$, and let its frequency be f_s Hz. The frequency of the sampling signal is also referred to as sampling frequency or sampling rate. By multiplying $x(t)$ and $y(t)$ we perform the impulse sampling, and the resultant is denoted as $z(t)$. From Fourier transform it is well known that such a multiplication operation in the time domain transforms into convolution in the frequency domain, as depicted in Figure 8.6.

Thus, it can be seen that for distortion-less sampling the key requirement is $f_s \geq 2 f_m$. This requires low-pass filtering of the analog signal prior to the sampling process with a cutoff of $(f_s/2)$ Hz. Otherwise, it would potentially

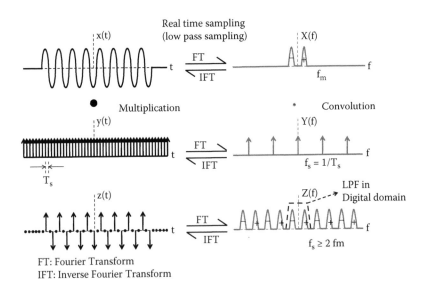

FT: Fourier Transform
IFT: Inverse Fourier Transform

FIGURE 8.6
Real-time oscilloscope: time and frequency domains.

result in aliasing in the frequency domain during sampling that cannot be recovered. Hence, such analog front-end low-pass filters have earned themselves a special name: antialiasing low-pass filter. Sampling of $x(t)$ results in $z(t)$. $z(t)$ is digitized to representation using a finite number of bits for each sample in the ADC. This is where quantization error enters the picture. But with higher bits leading to finer resolution of digitizer (ADC), the effect of quantization error can be reduced to acceptable levels. Now the digitized $z(t)$ is processed in the digital domain to extract various information, like maximum value, minimum value, period, rise time, delay time, frequency spectrum, etc. Before display, $z(t)$ (or processed digital signal) is low-pass filtered in the digital domain for better reconstruction of a continuous analog signal on the display. For example, this digital low-pass filtering can range from simple linear interpolation to well-defined sophisticated $\sin(x)/x$ interpolations and their variants for better reconstruction and smoother displays. This type of real-time sampling is well known as low-pass sampling in communication theory. Figure 8.6 gives a complete description pictorially.

Now what would happen in the presence of noise embedded in a signal. Let us make a qualitative analysis. Hence, let the noise be additive white Gaussian noise (AWGN) with zero mean. Let the power spectral density (PSD) of AWGN be N_0 dbm/Hz (single-sided), which is constant. Let the average signal power be S dBm. Now, one can define the signal-to-noise ratio (SNR) of the impulse sampler after low-pass filtering as the ratio of average signal power captured to the average noise power captured.

$$\text{SNR} = \frac{\text{average Signal power captured } (W)}{\text{average Noise power captured } (W)} \tag{8.1}$$

$$\text{SNR (dB)} = \text{average Signal power (dBm)} - \text{average Noise power (dBm)} \tag{8.2}$$

$$\text{SNR (dB)} = S - N_0 B \tag{8.3}$$

where B is the equivalent noise bandwidth of the antialiasing low-pass filter, prior to impulse sampling. One important observation is that for a given signal power, SNR of real-time sampling degrades with increase in bandwidth.

Example 8.3

Let an ideal antialiasing low-pass filter with bandwidth (B) of 20 GHz be used prior to impulse sampling. Let the sampling rate be 40 GSps. A 5 GHz sinusoidal signal with available power of 0 dBm is impulse sampled. The thermal noise PSD is given by $N_0 = kT$ (W/Hz), where k

is the Boltzmann constant $(1.38 \times 10^{-23} \text{ JK}^{-1})$ and T is the temperature in Kelvin (K). At 300 K:

$$\text{Total noise power captured } (N_0 B) = kTB$$

$$kTB = 1.38 \times 10^{-23} \times 300 \times 20 \times 10^9 = 8.28 \times 10^{-11} \text{ W} = -100.8 \text{ dBW}$$

$$\text{Total noise power captured } (N_0 B) = (-100.8 + 30) = -70.8 \text{ dBm}$$

$$\text{Total signal power captured } (S) = 0 \text{ dBm (given)}$$

Thus,

$$\text{SNR(dB)} = S - N_0 B = 0 - (-70.8) = 70.8 \text{ dB}$$

Example 8.4

Let an ideal antialiasing low-pass filter with bandwidth (B) of 40 GHz be used prior to impulse sampling. Let the sampling rate be 80 GSps. A 5 GHz sinusoidal signal with available power of 0 dBm is impulse sampled. The thermal noise PSD is given by $N_0 = kT$ (W/Hz), where k is the Boltzmann constant $(1.38 \times 10^{-23} \text{ JK}^{-1})$ and T is the temperature in Kelvin (K). At 300 K:

$$\text{Total noise power captured } (N_0 B) = kTB$$

$$kTB = 1.38 \times 10^{-23} \times 300 \times 40 \times 10^9 = 1.65 \times 10^{-10} \text{ W}$$

$$\text{Total noise power captured } (N_0 B) = -67.8 \text{ dBm}$$

$$\text{Total signal power captured } (S) = 0 \text{ dBm (given)}$$

Thus,

$$\text{SNR(dB)} = S - N_0 B = 0 - (-67.8) = 67.8 \text{ dB}$$

Thus, comparing Examples 8.3 and 8.4, we can see that when the same signal is captured in an idealized impulse sampler with twice higher bandwidth, the SNR is degraded by 3 dB. We shall explore more on noise as we move on, covering new topics. Also notice that we have ignored quantization noise.

Real-time oscilloscopes offer a super challenge when it comes to capturing signals at microwave frequency. For example, to acquire a K band carrier signal at 20 GHz, a real-time digital oscilloscope would require at least 40 GSps ADC and a memory depth of 1 Gpts to acquire 25 ms. However, the single most undisputed advantage is, as the name upholds, real-time oscilloscopes are indeed real. That means, real-time oscilloscopes capture the incoming signal as it really is, within the limitation of ADC sampling rate. One would appreciate this fact after going through the operating principle of sampling oscilloscope.

8.2.1 Sampling Oscilloscope (Equivalent Time Oscilloscope)

Sampling oscilloscopes cleverly use the repetitive property of the incoming signal to relax the requirement of ADC sampling rate. Basically, samples are acquired from many periods of the incoming signal. These samples are then interleaved to create one complete period. Figure 8.7 explains this in a much better way pictorially.

The key observation is that one sample is acquired for every trigger, and successive triggers are displaced with a well-defined time delay or time offset referred to as delta_time (δ) in Figure 8.7. One of the key requirements is to derive or obtain a trigger signal synchronized with the incoming repetitive signal. Notice that the triggering signal too is a periodic signal with period $(T_p + \delta)$, as seen in Figure 8.7. This operating principle is known as sequential sampling oscilloscope. There is another variety known as random sampling oscilloscope, which differs slightly with random sampling as opposed to fixed-delay trigger in a sequential sampling oscilloscope.

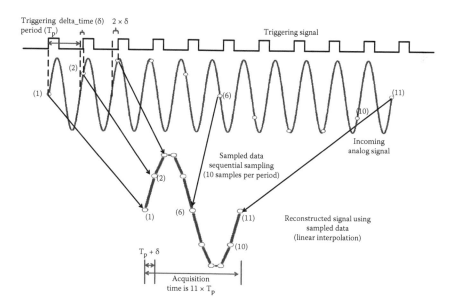

FIGURE 8.7
Sequential sampling oscilloscope: operating principle.

Example 8.5

Let a sequential digital sampling oscilloscope be used to capture one complete period of C band repetitive sinusoidal signal at 5 GHz with a time resolution of 0.02 ns using a synchronized triggering period of 0.01 μs (sampling rate of 100 MSps). This would require a total of 10 pulses of triggering signal with 10.02 ns interpulse spacing. Thus, the total data acquisition time is 0.1 μs (i.e., approximately 10×0.01 and exactly 0.1002 μs). On the other hand, a real-time oscilloscope would require an ADC with sampling rate of at least 50 GSps, to achieve the same time resolution of 0.02 ns.

As can be seen from Example 8.5, it is very evident that sequential sampling oscilloscopes greatly relax the ADC sampling rate for desired time resolution at microwave frequencies compared to real-time oscilloscopes. However, the price paid is that the incoming signal is required to be repetitive. Random spikes or dips will not only be lost, but can create a false impression of the signal. Hence, for capturing nonperiodic signals, a sequential sampling oscilloscope cannot be used, but a real-time oscilloscope with necessary sampling rate can be used (Agilent Technologies, 2013b; Tektronix, 2001). Thus, as previously mentioned, a real-time oscilloscope captures the incoming signal as it really is. Also important to notice is that the time offset denoted as delta_time in Figure 8.7 has to be generated with precision for the success of operation of the sequential sampling oscilloscope?

Example 8.6

Let a sequential digital sampling oscilloscope be used to capture one complete period of 5 GHz signal that is amplitude modulated (say, ON/OFF keying) by a 100 MHz square wave. The required time resolution is 0.02 ns using a synchronized triggering period of 0.01 μs (sampling rate of 100 MSps). Since the period of the incoming signal is now 10 ns, and the time resolution is 0.02 ns, it would require a total of 500 runs of triggering signal. Thus, the total data acquisition time is 5 μs (i.e., 500×0.01 μs).

Now let us see the frequency domain behavior of sequential sampling operation. Unlike real-time sampling (or low-pass sampling), we get a hint that this sequential sampling operation may not be that simple and obvious after all. Since we are sampling at a lesser sampling rate, we predict that the sampled spectrum gets aliased.

Figure 8.8 gives the description of sequential sampling of a continuous sinusoidal signal of period T_p denoted as $x(t)$. Notice that this time around the incoming signal spectrum denoted as $X(f)$ is impulsive due to the periodic nature, which is a must for the sampling oscilloscope. Let δ be the required time resolution. In Figure 8.8, $\delta = (T_p/4)$ s is assumed just for illustration purposes. Thus, the triggering signal (which is the sampling signal) represented

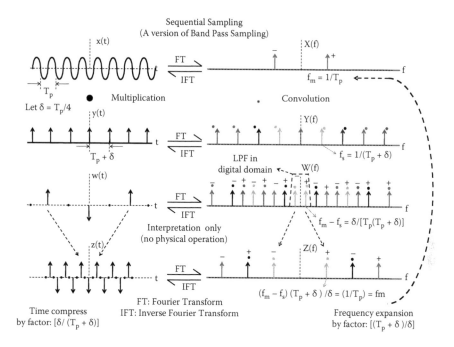

FIGURE 8.8
Sequential sampling oscilloscope: time and frequency domains.

as $y(t)$ has the period of $(T_p + \delta)$. Thus, we are sampling at a rate $\{1/(T_p + \delta)\}$ Sps (samples per second), which is less than the incoming signal frequency $(1/T_p)$ Hz. $Y(f)$ shows the spectrum of $y(t)$. For better visualization, color mapping is used for different harmonics. This color mapping becomes very helpful as we move on. The sampled signals at low data rate denoted by $w(t)$ are then placed side by side. The key point is that now they are interpreted to be spaced by intervals of δ instead. That is, after sampling, the sampling period is intentionally interpreted as δ, even though in actuality it is known to be $(T_p + \delta)$ seconds. The signal obtained after this new interpretation is denoted by $z(t)$. This is equivalent to time compression of sampled data, which corresponds to expansion in the frequency domain. The time compression factor is $[\delta/(T_p + \delta)]$, because $(T_p + \delta)$ now corresponds to δ. Thus, the frequency expansion factor is $[(T_p + \delta)/\delta]$ from Fourier transform. Note that during time compression or frequency expansion operation, there is no change in the power since we are handling the periodic signals. This interpretation perfectly places the low data rate sampled signal at the place of original spectrum. Thus, the sampled and stretched spectrum $Z(f)$ captures full information of the periodic incoming spectrum $X(f)$, as depicted in Figure 8.8.

Now digitized $z(t)$ is then processed in the digital domain to extract the required information. Before display, $z(t)$ (or processed digital signal) is low-pass filtered in the digital domain for better reconstruction of the continuous analog signal on the display. This type of sequential sampling is very much similar to the well-known band-pass sampling. However, there are many versions and variations in performing band-pass sampling. One can consider sequential sampling as one such variation of band-pass sampling, which is applicable for periodic signal only.

One interesting observation is that unlike real-time sampling discussed in the previous section, sequential sampling is suited for time domain. That is, sequential sampling is better analyzed in the time domain than the frequency domain. For example, from the incoming signal spectrum it is very comfortable to determine the sampling rate requirement in case of real-time sampling. In fact, the key design parameter in real-time sampling is that the sampling rate should be twice the highest frequency of interest. However, in sequential sampling it is tough to even determine the fundamental period of the incoming signal from the spectrum. The sampling rate in sequential sampling seems to be more governed by ability to generate precise time offset (δ) in the time domain rather than the incoming frequency spectrum. Also, another key requirement in sequential sampling, as discussed earlier, is the generation of triggering pulses synchronized with the incoming periodic signal. Again, this has to directly deal with time domain features.

For the specific example shown is Figure 8.8, let us analyze both aliasing condition and noise behavior. From the spectrum of $W(f)$ in Figure 8.8, we see that the signal of interest is generated by fundamental frequency content of $y(t)$ for this specific case of sequential sampling. Remember that for real-time sampling it is the content around dc that generates the signal of interest, and the fundamental frequency content causes aliasing. However, in sequential sampling of Figure 8.8, from $W(f)$ we also determine that the nearest possible aliasing can be caused by the second harmonic. Thus, the nonaliasing condition (i.e., to prevent aliasing) as applicable for the specific example of sequential sampling shown in Figure 8.8 is

$$(2f_s - f_m) \geq (f_m - f_s)$$

$$3f_s \geq 2f_m \tag{8.4}$$

Let us further express this equation in terms of time domain parameters, i.e., δ (time offset) and T_p, as below:

$$\frac{3}{T_p +} \geq \frac{2}{T_p}$$

$$T_p \geq \frac{2}{3} \tag{8.5}$$

In general, sequential sampling oscilloscopes have a fixed value of δ. Thus, for the continuous sinusoidal signal case, as considered in Figure 8.8, this would require an antialiasing low-pass filter (LPF) whose cutoff is $1.5/\delta$, as derived above.

Noise analysis is a bit more involved than the real-time sampling case. Hence, let us simplify the assumptions to obtain a gross but useful conclusion. From Figure 8.8, we make two important observations. First is that the bandwidth of acquisition is less. Thus, it may initially appear that the noise captured is less. But there is a second observation. As seen from spectrum $W(f)$ in Figure 8.8, noise gets added at the acquisition band because of aliasing from multiple harmonics of the sampling signal. In order have a firsthand feel of noise behavior, let us take the simplest case. As in the case of real-time sampling, let us assume additive white Gaussian noise (AWGN) with zero mean. Let $N_0 = kT$ (W per Hz) be the PSD of the noise (single-side spectrum, so if negative frequencies are counted, the PSD would be $N_0/2$). Let the frequency of input periodic sinusoidal signal be f_m Hz. Thus, the period of the signal is $(1/f_m)$ seconds. Let this be passed through an ideal LPF of cutoff slightly greater than f_m Hz to prevent aliasing prior to sampling. Thus, we can assume that noise outside the frequency f_m of the incoming signal passed through the aliasing filter is zero. This is one of the key assumptions. Now for the case depicted in Figure 8.8, we see that only dc and first harmonic mixing bring noise from three regions of the spectrum (around dc, around f_m, and around $-f_m$) to the captured bandwidth, which is denoted by $-B$ to B Hz (that is, $-(f_m - f_s)$ to $(f_m - f_s)$). Thus, the total noise captured is calculated as follows: dc mixing results in noise power of kTB. The first harmonic mixing results in noise power of $2\,kTB$. The total noise power captured $= 3\,kTB$. Now if signal power acquired is S watts, then SNR sequential sampling as applicable to Figure 8.8 under the assumptions mentioned is given by $S/3$ kTB in value.

Example 8.7

Referring to the sequential sampling shown in Figure 8.8, let the captured bandwidth post sampling be $(f_m - f_s)$ Hz. As seen in Figure 8.8, δ is $(T_p/4)$.

Now the total noise captured is $3\,kTB$ watts, but $B = f_m - f_s$:

$$B = \delta/[(T_p)\,(T_p + \delta)]$$

Now substituting $\delta = (T_p/4)$,

$$B = 1/(5\,T_p)$$

Thus, if the signal power acquired is S watts, then $SNR = (5\,ST_p)/(3\,kT)$.

Example 8.8

Let us compare the results from Example 8.7, with real-time sampling for the same assumptions.

Real-time sampling:

$$\text{Noise power} = kTf_m \text{ watts} = kT/T_p \text{ watts}$$

Thus,

$$\text{SNR}_{real\_time} = ST_p/kT$$

Sequential sampling:

$$\text{SNR}_{sequential} = (5\ ST_p)/(3\ kT)$$

$$\text{SNR}_{sequential}/\text{SNR}_{real\_time} = (5/3)$$

$$\text{SNR}_{sequential}\ (\text{dB}) = \text{SNR}_{real\_time}\ (\text{dB}) + 2.22\ \text{dB}$$

Thus, it seems that $\text{SNR}_{sequential}$ (dB) is better than SNR_{real_time} by 2.22 dB. Note that this is applicable only for specific case of Figure 8.8, under all the simplifying assumptions.

Actually, for most practical cases we can assume that the SNRs of both real-time sampling and sequential sampling are almost the same.

Let us summarize the real-time oscilloscope and sampling oscilloscope in Table 8.2.

Whether is it a real-time oscilloscope or sampling oscilloscope, it is evident that the ADC is the heart of any digital oscilloscope. The higher the sampling rate of ADC, the higher is the frequency that can be captured in the oscilloscope. Hence, let us explore more on high-speed ADC in the following section.

TABLE 8.2

Comparison Table between Real-Time Oscilloscope and Sampling Oscilloscope

Real-Time Oscilloscope	Sampling Oscilloscope
Equivalent to low-pass sampling	Equivalent to band-pass sampling
ADC with extremely large sampling Rate is required for sampling Microwave signal	ADC sampling rate is relaxed, but very fine control over the sampling interval is required
Resolution (number of bits) tends to be lower as sampling rate increases	Resolution is much higher than for real-time oscilloscope
Can be used on either periodic or nonperiodic signal	Strictly requires periodic signal

8.3 High-Speed ADC: Working Principle

An analog-to-digital convertor (ADC) is the key component for successful operation of any digital oscilloscope. The emphasis gets enhanced with microwave oscilloscopes. Even though, theoretically, a sampling rate of twice the highest frequency is sufficient from the Nyquist theorem, in reality it is much higher. For example, Figure 8.9 shows an analog sinusoid signal being reconstructed from sampled data using simple linear interpolation. It is very evident that around 10 samples are required to have good integrity visually. Thus, for this particular case, the sampling rate needed to capture, say, a 5 GHz signal is around 50 GSps. A rule of thumb of five samples per period is used in most applications. Some oscilloscopes reduce this number to as low as 2.5 samples per period (approaching the Nyquist criterion) using a sophisticated interpolation algorithm like $\sin(x)/x$. Can we get an ADC at such a high sampling rate? What would be its architecture? These are some of the questions that arise soon.

ADC is also referred to as digitize because it performs both sampling and digitization. The incoming signal is multiplied by the sampling signal, which

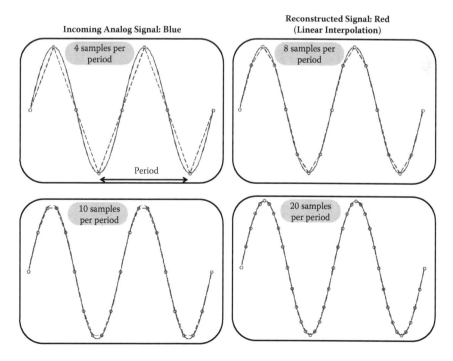

FIGURE 8.9
Example to show reconstruction.

is a train of impulses with a sampling interval of T_s seconds. The inverse of T_s is known as the sampling rate or sampling frequency F_s Sps (samples per second), i.e., $F_s = 1/T_s$. In the frequency domain this would lead to additional repetition of the spectrum of incoming signal at F_s and its harmonics. Note that for preventing aliasing, the sampling rate should be at least twice the highest frequency of the incoming signal. In other words, the incoming signal needs to be filtered using antialiasing LPF with cutoff around $F_s/2$, to prevent an aliasing effect. Note that at the end of impulse sampling, the signal is still analog but discrete.

Quantization now has to be performed on the sampled analog signal. Quantization is required because the discrete signal needs to be represented using a finite number of bits for either digital processing or digital storing. This leads to quantization error, which cannot be removed. However, by using a sufficient number of bits, the quantization error can be minimized. Figure 8.10 gives the pictorial representation of the quantization process in a typical ADC.

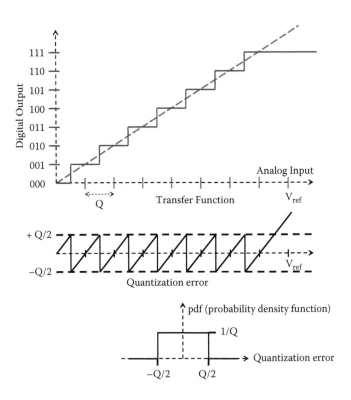

FIGURE 8.10
Quantization process.

Figure 8.10 describes the quantization process for a 3-bit ADC. Both transfer function and quantization error are described. As can be seen, the quantization error cannot be nullified because there is no one-to-one relation between the incoming analog signal and quantization error. Q is referred to as a quantization bin. For an N-bit ADC (N digital output bits or N levels) with a reference of V_{ref}, let us determine what is the level of quantization error on the signal quality.

$$Q = \frac{V_{ref}}{2^N} \tag{8.6}$$

Assuming uniform probability density function (pdf) for the quantization error as shown in Figure 8.10, the variance (σ_{QE}^2) can be shown to be

$$\sigma_{QE}^2 = \frac{Q^2}{12} \tag{8.7}$$

Assuming an input sinusoidal with peak-to-peak amplitude of V_{ref}, the rms voltage is given by

$$V_{RMS} = \frac{V_{ref}}{2\sqrt{2}} = \frac{Q\,2^N}{2\sqrt{2}} \tag{8.8}$$

Thus, by defining the signal-to-quantization error ratio (SQR) for an N-bit ADC as the ratio of input signal power to the quantization error power, we get

$$SQR = \frac{\text{Signal Power}}{\text{Quantization Error Power}} \tag{8.9}$$

$$SQR = \frac{V_{RMS}^2}{\sigma_{QE}^2} \tag{8.10}$$

$$SQR\ (dB) = 20\log\left(\frac{V_{RMS}}{QE}\right) \tag{8.11}$$

$$SQR\ (dB) = 20\log\left(\frac{2^N\sqrt{12}}{2\sqrt{2}}\right) \tag{8.12}$$

$$SQR\ (dB) = 6.02\ N + 1.76 \tag{8.13}$$

This is a familiar equation widely found in literature (Kester; Bowling, 2000). One important observation is that by using ADC with larger bits, i.e., N, we can increase the SQR for a given incoming signal, and thus minimize the quantization error to a lower and at times acceptable level. This is very evident physically, because with a larger value of N, now V_{ref} is divided into many more levels, thus reducing the quantization error. Note that in the calculation of SQR above, the incoming signal is assumed to be noise-free.

Example 8.9

A real-time digital oscilloscope has a sampling rate of 2 GSps and resolution of 16 bits. Thus, the memory depth required to capture 1 s duration of a signal is 32 Gbps (gigabits per second) or, equivalently, 4 GBps (gigabytes per second). SQR of the ADC is 6.02 (16) + 1.76 = 98.08 dB.

8.3.1 Flash ADC Architecture

The fastest ADC architecture is known to be Flash ADC or direct conversion ADC (Ahmed, 2010). It basically uses a parallel conversion methodology. The input signal is fed to multiple comparators in parallel followed by a priority encoder to generate the required digital bit pattern. Figure 8.11 depicts the typical architecture of a 3-bit Flash-based ADC.

Example 8.10

Consider the 3-bit Flash ADC depicted in Figure 8.11. If $V_{ref} = 5$ V and $V_{in} = 2.4$ V, then the comparators set to high level, i.e., 1 ($\cong V_{dd}$), are shown in Figure 8.12. The priority encoder would give the priority to bit position 3 and correspondingly encode it to 011.

Theoretically, a single clock cycle is sufficient for analog-to-digital conversion in Flash ADC. This is the fastest electronic ADC architecture being used commercially. But the speed comes at a price. As seen from Figure 8.12, even for 3-bit resolution one requires around seven comparators and an 8:3 priority encoder. Thus, an ADC with 8-bit resolution would require 255 comparators and a 256:8 priority encoder. Thus, the resource, chip area, and power would increase considerably with every addition of a bit. Thus, Flash ADCs operating at few GSps usually have less resolution, like 8, 10, or 12 bits of resolution. With the tremendous development in technology, the sampling speed of Flash ADC is ever increasing. Today, one can purchase 4 GSps with 12-bit resolution. This ADC would be sufficient to capture and analyze signals up to around 1 GHz.

However, one can find commercial real-time microwave oscilloscopes that claim to have sampling rate ranging from 40 to 80 GSps and beyond. So, where do these ultra-high-speed ADCs come from? Having a real-time

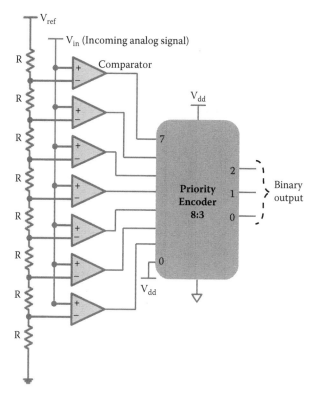

FIGURE 8.11
Architecture: 3-bit Flash ADC.

digital oscilloscope that enables one to view a K band signal at 25 GHz not only gives immense pleasure, but also empowers the radio frequency (RF) and microwave engineer with a tool, which a few decades ago was thought next to impossible. So, what made this possible today?

Let us explore some of the innovative mechanisms that are being used in today's state-of-the-art real-time microwave oscilloscopes, whose ADC sampling has reached hundreds of GSps (up to 160 GSps today), and is fast increasing as days pass. They are:

1. Interleaved ADCs: Electronic ADC architecture found in most oscilloscopes.
2. Photonic ADC: Waiting to get commercialized in the near future.

8.3.2 Interleaved ADCs

The idea of interleaved ADCs is nothing but putting many ADCs in parallel to enhance the sampling rate. It is like when one person is not able to

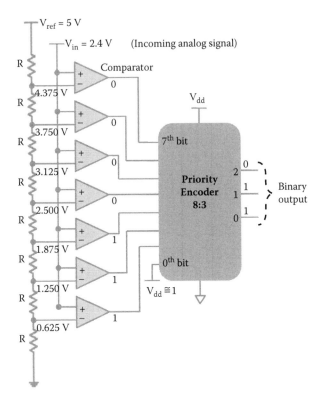

FIGURE 8.12
Circuit for Example 8.10.

complete a job quick enough, so many are put in placed and asked to work in parallel to get the job done quickly. Most modern available real-time oscilloscopes use this principle effectively. They actually have tens of Flash ADCs operating with well-defined time offsets to capture signals at tremendous sampling rates like 20 GSps and beyond. The principle is that simple. For example, suppose each ADC has a sampling rate of 2 GSps and resolution of 12 bits. Now by connecting 20 such ADCs in parallel with a suitable time offset, one can have the ability to sample the incoming signal at 40 GSps with 12-bit resolution. Yes, indeed, this is what has made microwave oscilloscopes a reality today. Figure 8.13 explains the time interleaving with five ADCs connected in parallel.

Example 8.11

The goal is to achieve a sampling rate of 20 GSps digitizer using an ADC with a sampling rate of 500 MSps. Thus, a total of at least 40 such ADCs are required. The time offset (shown in Figure 8.13) required is 50 ps.

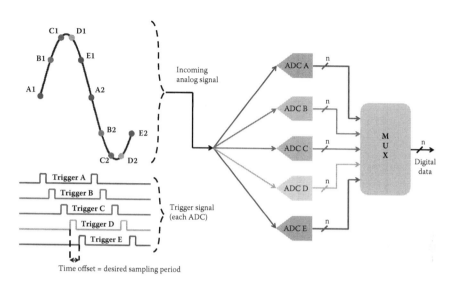

FIGURE 8.13
Interleaved ADCs.

However, there are practical challenges. Parallel configuration, of course, increases the overall sampling ability. But it also increases ADC coupling (operation of one ADC affecting the signal input to another ADC). Particularly at microwave frequency, the inter-ADC coupling or loading becomes difficult to handle. It requires sophisticated hardware as well as software solutions to address the issue of coupling among parallel ADCs. In addition, when time interleaving enters the picosecond range, timing accuracy and precision become the prime focus and present considerable system challenges. Hence, it is very desirable to reduce the number of ADCs.

Commercially available ADCs have a maximum sampling of around 4 GSps. But oscilloscope manufacturers don't use these commercially available ADCs. They have their own versions of ultra-high-speed ADCs. In fact, companies keep their ADC architecture secret because of its commercial viability. Another requirement for such high-speed ADCs is that their analog front-end bandwidth also needs to be high, at least up to half the sampling frequency. This bandwidth may easily extend to 20 or 30 GHz, or even more as sampling rate increases. The traditionally preferred Si substrate cannot achieve such high bandwidth due to lower electron mobility. There have been many significant research and developmental activities in recent years to address the issue of high bandwidth and high data rates. This eventually has led to innovations on materials, circuits, and packaging. New substrates like SiGe and InP have jumped from labs to products. These new substrates have significantly higher electron mobility, and hence significantly larger bandwidth, than Si substrates. From the circuit point of view, the transistor

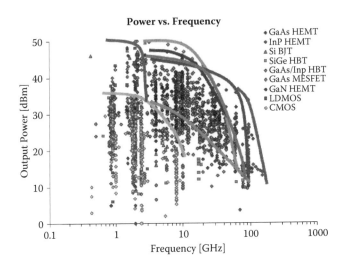

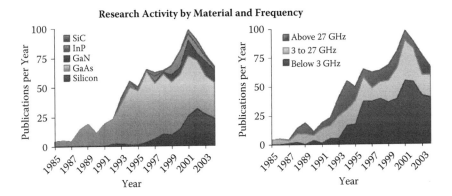

FIGURE 8.14
Comparison of technologies and research scenario.

has gone through tremendous variations like heterojunction bipolar (HBT) and high-electron-mobility (HEMT) transistor with increased bandwidth. Figure 8.14 (del Alamo, 2005) makes a qualitative comparison among different transistor technologies.

Figure 8.14 clearly shows that SiGe and InP IC can and have achieved bandwidths well above 20 GHz, almost reaching 100 GHz. One very interesting observation from the graph of research activity is that the 1990s is when we started to see the surge of publications in SiGe and InP technology. However, the trend seems to be reducing after 2003. This is where the research ideas transformed into today's products.

As an example, as of today, high-bandwidth oscilloscopes from Tektronix use commercial SiGe for their preamp circuits (Tektronix, 2013a), whereas

FIGURE 8.15
Multichip module used in Agilent Infiniium high-speed real-time oscilloscopes.

Teledyne LeCroy and Agilent use an in-house proprietary InP process (Teledyne LeCroy, 2013; Agilent Technologies, 2013c). The multichip module shown in Figure 8.15 is said to be used in high-speed real-time oscilloscopes from Agilent, whose sampling rate has reached 160 GSps as of today, and is expected to keep rising. This multichip module is quoted to have an analog bandwidth of 33 GHz. With a couple of such ultra-high-speed ADCs and using the interleaved architecture, real-time oscilloscopes can achieve hundreds of GSps.

8.3.3 Photonic ADC

The principle of photonic ADC is relatively recent. The idea was proposed by Bahram Jalali et al. in 1998 (Bhushan et al., 1998; Jalali et al., 1998; Coppinger et al., 1999). So far, we have discussed electronic ADCs at length. These electronic ADCs operate purely in the electrical domain and are limited in their operating speed. The limitation has been that the sampling rate of a single ADC is not fast enough to capture a microwave signal. Hence, we have seen how time interleaving and parallel architecture have come to the rescue. But the system gets too complicated and sophisticated. So, in the quest to find a solution where a single ADC could be used to capture even a microwave signal in real-time sampling, the answer is found in the optical domain.

Here is the idea: if sampling period is not small enough, then stretch the input signal and then sample at the available sampling speed. This will effectively reduce the sampling rate required. Figure 8.16 explains this pictorially. Theoretically, if the incoming signal is scaled up (stretched) by a factor

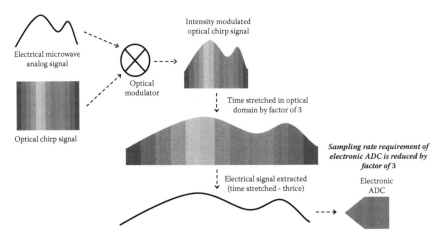

FIGURE 8.16
Timescaling principle of photonic ADC.

N in the time domain, then the signal gets scaled down (compressed) by the same factor N in the frequency domain from Fourier transform. Thus, the required sampling rate is also reduced by the same scaling factor N. With the knowledge of value of N, the complete information is recovered after such sampling.

Example 8.12

With a timescaling factor (stretching factor) of 20, one can use a 4 GSps ADC to act like an 80 GSps ADC.

So, the question now is: How do we time stretch the incoming microwave signal? Figure 8.17 shows the block-level diagram (Coppinger et al., 1999)

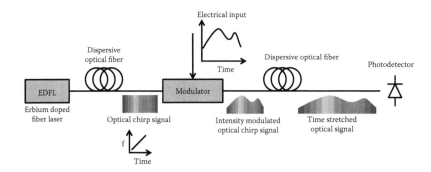

FIGURE 8.17
Block diagram of time stretching.

TABLE 8.3

Stretch Factor as Function of Fiber Length

Length of Optical Fiber (km)	Stretch Factor (or Scale Factor)
2.2	3
5.5	6
7.6	8

explaining the time stretching performed in the optical domain. A chirp signal is one whose frequency varies with time, as shown by color variation in Figure 8.17 (when birds chirp, they produce sound signals with time-varying instantaneous frequency). This optical chirp signal is amplitude modulated with the incoming electrical microwave signal to be sampled. The modulated optical signal is then time stretched. A simple optical fiber with its dispersive behavior comes in handy to time stretch the optical signals. This is where the optical domain comes to the rescue, to time stretch the signal. When the modulated optical signal is time stretched, so is the envelope (carrying the information to be sampled). Finally, the envelope is recovered in the electrical domain, which now has a frequency down by the stretch factor. This low-frequency electrical signal is sampled with a commercially available electronic ADC of suitable sampling rate, as shown in Figure 8.17. Thus, with the knowledge of stretch factor, one can achieve an ultra-high sampling rate of up to hundreds of GSps.

One important observation is that the stretch factor determines the frequency downscaling. For a given type of dispersive optical fiber, it is observed that the stretch factor increases with increase in optical length. Table 8.3 gives a feel for this dependency [16].

As of today, all commercial oscilloscopes use electronic ADC. Hopefully in the near future one may find commercial oscilloscopes that use the photonic stretching described above.

8.4 Limitations of the Real-Time Oscilloscope

From discussion in the previous sections, we understand that the real-time microwave oscilloscope is one of the most versatile measurement equipment. Of the two types of microwave oscilloscopes, real-time oscilloscopes can characterize both periodic and nonperiodic signals, unlike sampling oscilloscopes, which can characterize only periodic signals. The utility of the real-time oscilloscope lies not only in characterizing microwave signals, but also in characterizing linear time-varying systems (like mixers or modulators). In addition, the real-time microwave oscilloscope becomes an invaluable instrument for characterizing nonlinear time-invariant systems, for a specific input. For example, in characterizing switching systems like the ASK modulator for switching response, the real-time microwave oscilloscope is a wonderful instrument with its ability to extract the time domain data for further signal processing. So, with real-time microwave oscilloscopes reaching sampling rates of hundreds of GSps, there follows a logical question without much delay. And that is: If a real-time microwave oscilloscope has such wonderful capability to characterize signals and systems, then why don't we replace the VNA or spectrum analyzer (SA) with it? Why should one buy both a real-time microwave oscilloscope and VNA when a real-time microwave oscilloscope can do the job? This is a very valid question, particularly in industry, where cost cutting is often as good as revenue. So, is a real-time microwave oscilloscope alone sufficient for complete microwave measurements? Do real-time microwave oscilloscopes have any limitations?

From an initial gross view, it seems that there is a severe limitation of real-time microwave oscilloscopes. That limitation comes from the fact that real-time microwave oscilloscopes are wide-band systems. And wide-band systems tend to be more noisy because the higher the bandwidth, the more the noise. VNA and SA have much lower bandwidth since they characterize the system or signal, frequency by frequency. They use mixing and measure the signal at intermediate frequency with a narrowband filter, and hence generate less noise. Thus, we see that VNA or SA tends to have lower noise than the real-time microwave oscilloscope (Agilent Technologies, 2010). Let us explore more on this limitation of real-time microwave oscilloscopes inherent to their wide-band nature in the following section.

8.4.1 Trade-Off: Bandwidth vs. Sensitivity in ADC

Real-time microwave oscilloscopes have very high sampling rates, extending to tens and hundreds of GSps. As we have seen in real-time sampling, to avoid aliasing, an analog antialiasing low-pass filter is used prior to sampling. Ideally, the oscilloscope should support high-analog bandwidth at the front end, which then is limited by this antialiasing low-pass filter, which has a cutoff of half the sampling rate (f_s). If the sampling rate is 100 GSps, then ideally the antialiasing low-pass filter should have a cutoff at 50 GHz. Thus, real-time microwave oscilloscopes are inherently wide-band systems. As a result, the total noise captured by the system is also high. The noise power captured for the case of additive Gaussian white noise with zero mean (like thermal noise) is kTB watts, where k is the Boltzmann constant (1.38×10^{-23} J/K/Hz), T is the temperature in Kelvin, and B is the system bandwidth. Thus, the higher the bandwidth, the higher is the noise power captured. This will reduce the ability of the oscilloscope to detect weak signals (sensitivity) because the weak signals get engulfed by the higher noise captured because of the wide bandwidth. Note that sensitivity is inversely related to minimum

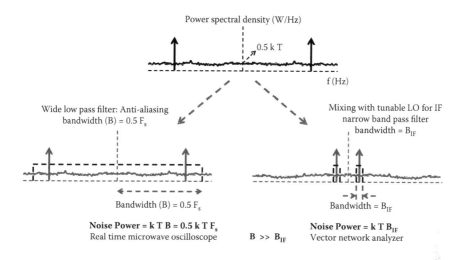

FIGURE 8.18
Real-time oscilloscopes vs. VNA: noise performance.

detectable power. Higher sensitivity means that very low input power levels can be detected. Thus, we see a trade-off between bandwidth and sensitivity in real-time microwave oscilloscopes. Sensitivity of the oscilloscope reduces with increase in its bandwidth. That is, the higher the sampling rate in real-time oscilloscopes, the lower its sensitivity.

Figure 8.18 depicts the noise power captured in the real-time microwave oscilloscope compared to VNA or the spectrum analyzer (SA). Unlike the oscilloscope, VNA or SA first downconverts the incoming signal to a fixed intermediate frequency (IF) by mixing with tunable local oscillator (LO) by rejecting the image frequency at the input. Then the IF is filtered using a narrow band-pass filter of bandwidth, say, B_{IF}. Then the band-pass-filtered IF signal is fed to ADC for further processing. Thus, as seen from Figure 8.18, the noise power at the input of ADC in VNA or SA is kTB_{IF}, which is much less than that of the real-time microwave oscilloscope.

The above phenomenon can be captured quantitatively using a parameter known as the dynamic range. The dynamic range of an instrument is defined as the ratio of maximum power detected to the minimum power that can be detected. For our case, let us define the dynamic range prior to ADC for both the real-time microwave oscilloscope and VNA or SA, thus removing the ADC dependency.

$$\text{Dynamic Range (DR)} = \frac{\text{Maximum Power Detected } (P_{max}) \text{ in watt}}{\text{Minimum Power Detected } (P_{min}) \text{ in watt}} \qquad (8.14)$$

$$\text{DR (dB)} = P_{max}(\text{dBm}) - P_{min}(\text{dBm}) \qquad (8.15)$$

The minimum power (P_{min}) that can be detected is equal to the average noise power (N) captured. If the signal power is less than the average noise power (N), then it cannot be distinguished from noise, and hence is lost.

$$\text{Minimum Power Detected } (P_{min}) = \text{Average Noise Power } (N) \qquad (8.16)$$

For the real-time oscilloscope, the average noise power prior to ADC sampling is

$$N = k\, T\, B = \frac{k\, T F_s}{2} \qquad (8.17)$$

where B is the antialiasing low-pass filter cutoff frequency. Ideally, B is half the sampling rate, i.e., $F_s/2$. Thus, for a real-time microwave oscilloscope we get

$$\text{Real Time Microwave Oscilloscope: } P_{min} = k\, T\, B = \frac{k\, T F_s}{2} \qquad (8.18)$$

$$DR = \frac{P_{max}}{P_{min}} = \frac{P_{max}}{k\, T\, B} = \frac{2\, P_{max}}{k\, T F_s} \qquad (8.19)$$

On the other hand, for VNA or SA, let the bandwidth of a narrow bandpass filter be B_{IF}. Also, let us assume that the mixing in VNA or SA with LO is ideal, so that no additional noise is added during mixing. In that case we get

$$N = k\, T\, B_{IF} \qquad (8.20)$$

$$\text{VNA or SA: } P_{min} = k\, T\, B_{IF} \qquad (8.21)$$

$$DR = \frac{P_{max}}{P_{min}} = \frac{P_{max}}{k\, T\, B_{IF}} \qquad (8.22)$$

For the sake of comparison, let us assume that the maximum power detected is the same for both types of instrument, i.e., for real-time oscilloscope and VNA or SA. Thus, comparing the DR (dynamic range) of both instruments we get

$$\frac{DR_{VNA}}{DR_{Osci}} = \frac{B}{B_{IF}} = \frac{F_s}{2\, B_{IF}} \qquad (8.23)$$

Since B_{IF} is much much lower than B, we clearly see that the dynamic range of VNA is much larger than that of the real-time oscilloscope. Thus, with increase in bandwidth of oscilloscope, the sensitivity reduces and as does the dynamic range.

$$B \gg B_{IF} \tag{8.24}$$

$$DR_{VNA} \gg DR_{Osci} \tag{8.25}$$

Example 8.13

If the intermediate frequency bandwidth (B_{IF}) of VNA is 1 MHz and the bandwidth (B) of a real-time oscilloscope is 40 GHz, then $DR_{VNA} = 4 \times 10^4 DR_{Osci}$. In other words, the dynamic range of VNA is improved by a factor of 46 dB compared to that of a real-time oscilloscope.

The summary is that SA or VNA can characterize weak signals or systems excited with weak signals. The higher the bandwidth of the oscilloscope, the lower will be its sensitivity. The average noise power at the input of ADC is directly proportional to the bandwidth, and hence the sampling rate. Thus, a real-time microwave oscilloscope is inferior to the VNA or SA in characterizing receiver systems, including an LNA (low noise amplifier). A receiver front end has very low input power levels in actual applications.

8.4.2 Trade-Off: Bandwidth vs. Sensitivity in Interleaved ADCs

In the previous section, we saw the trade-off between bandwidth and sensitivity in real-time microwave oscilloscopes using a single ADC. In this section, let us see how the interleaved ADCs explained in Figure 8.13, repeated below in Figure 8.19, behave with respect to the trade-off. The trade-off is initiated due to coupling of power to the ADCs that are OFF. That is, in Figure 8.19, at the instant when ADC A is ON (sampling), ideally no power should leak to other ADCs. But in practice, the power leaks because of imperfect isolation. Let this average power leaked to the ADC that is in OFF state be referred to as coupling (C).

$$\text{Coupling } (C) = \frac{\text{Power Leaked } (P_{leak}) \text{ in watt}}{\text{Input Power } (P_{in}) \text{ in watt}} \tag{8.26}$$

$$C(\text{dB}) = P_{leak}(\text{dBm}) - P_{in}(\text{dBm}) \tag{8.27}$$

Let N be the number of identical ADCs connected in parallel, each having a sampling rate of F_s Sps. Let P_{ADC_min} be the minimum input power that the ADCs can distinguish from noise. Note that sensitivity is inversely related to minimum detected power. Higher sensitivity means that ADC can detect very low input power levels. Let C be the coupling to each ADC that is OFF.

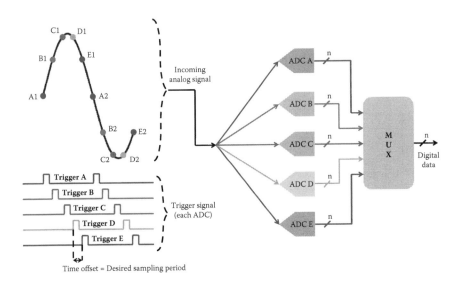

FIGURE 8.19
Interleaved ADCs.

It is assumed that ADC has sufficient analog bandwidth. Then the overall sampling ability of the interleaved ADC structure is N times F_s Sps.

$$\text{ADC sampling} = F_s \qquad (8.28)$$

$$\text{System Sampling} = N \ F_s \qquad (8.29)$$

The expression for input power (P_{in}) is

$$P_{in} = \text{ADC sensitivity} + (N-1) \text{ power leaked to ADC} \qquad (8.30)$$

$$P_{in} = P_{ADC\_min} + (N-1) \ C \ P_{in} \qquad (8.31)$$

Thus, the minimum input power required now for the interleaved system is

$$\text{Minimum Power required} = \frac{P_{ADC\_min}}{1-(N-1)C} \qquad (8.32)$$

$$P_{sys\_min} = \frac{P_{ADC\_min}}{1-(N-1)C} \qquad (8.33)$$

We see that the denominator is a positive quantity but less that 1. Thus, the minimum power required to detect the signal in the interleaved system is higher than that required for a single ADC. Thus, the system sensitivity is reduced in the course of increasing the sampling rate. Examples 8.14 and 8.15 clarify this.

Example 8.14

Let five identical ADCs be interleaved to enhance the sampling rate. Let each ADC have a sampling rate (F_s) of 4 GSps. Let the minimum power required for the ADC be 0 dBm. Let the coupling be −10 dB; i.e., 10% of input power is leaked to each ADC that is OFF. Now by interleaving five ADCs, the overall sampling rate is increased to 5×4 GSps = 20 GSps. But then the minimum power required for the system now is 2.22 dBm (i.e., $P_{sys\_min} = P_{ADC\_min} - 10 \log(1 - 4{*}0.1)$).

Example 8.15

Let five identical ADCs be interleaved to enhance the sampling rate. Let each ADC have a sampling rate (F_s) of 4 GSps. Let the minimum power required for the ADC be 0 dBm. Let the coupling be −15 dB; i.e., 3.2% of input power is leaked to each ADC that is OFF. Now by interleaving five ADCs, the overall sampling rate is increased to 5×4 GSps = 20 GSps. But then the minimum power required for the system now is 0.59 dBm (i.e., $P_{sys\_min} = P_{ADC\_min} - 10 \log(1 - 4{*}0.01)$).

8.4.3 Trade-Off: Bandwidth vs. Sensitivity in Photonic ADC

In this section let us have a look at how the trade-off between bandwidth and sensitivity expresses itself in photonic ADC. Photonic ADC, as mentioned earlier, is not used in any commercial microwave oscilloscope as of now. However, we have seen how the technology of InP and SiGe transformed from ideas in publication to today's high-end ADC, in just a decade. Figure 8.17, which is repeated as Figure 8.20, describes the block diagram of a typical photonic ADC. As discussed there, the key operation in photonic ADC is to time stretch the signal in the optical domain. Then, it is sampled using commercially available electronic ADC. Time stretching leads to frequency compression (downscaling), and hence reduces the need for ultra-high-speed electronic ADC. This becomes extremely vital when the bandwidth of an oscilloscope enters into hundres of GHz.

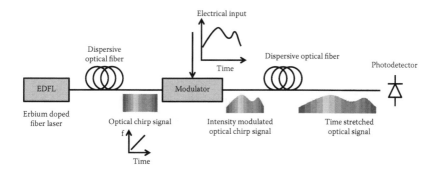

FIGURE 8.20
Photonic ADC block diagram.

The time domain stretch factor is an important parameter of photonic ADC. For a given electronic ADC being used, the overall sampling rate, and hence the bandwidth of photonic ADC, is directly proportional to the stretch factor. That is, double the stretch factor, and the effective sampling rate of photonic ADC doubles. Time domain stretching is performed in the optical domain using dispersive optical fibers. Referring to Figure 8.20, the optical fiber following the modulator performs the required time stretching. As the length of the dispersive optical fiber increases, so does the stretch factor. Thus, by making the dispersive optical fiber longer and longer, one can increase the time stretch factor, and thus the effective sampling rate and bandwidth. Ideally, this is perfect, but in practice, the dispersive optical fibers have loss. This loss in optical fiber is what is responsible for the trade-off. The loss or attenuation in optical fiber apart from reducing the signal strength of the modulated optical signal also adds noise (thermal loss gives thermal noise). Using a very low noise optical pump, one can account for attenuation in modulated signal strength. But the noise added due to attenuation or fiber loss is what reduces the sensitivity. Because of increase in noise power due to attenuation in longer fiber, the minimum power that can be detected increases, thus reducing the sensitivity. Thus, we can see the trade-off between bandwidth and sensitivity in photonic ADC as well.

Now that we have explored one of the important limitations of microwave oscilloscopes, it is time to update our Table 8.1 comparison with VNA and SA. We have seen that due to wide bandwidth, the real-time microwave oscilloscopes have lower sensitivity. The noise power is very high compared to VNA or SA. VNA or SA uses a narrow band-pass filter around intermediate frequency, thus capturing less noise. Thus, sensitivity of microwave oscilloscopes is much less than that of VNA or SA, particularly at microwave frequency. Table 8.4 is an updated version of Table 8.1. Note that sensitivity is the ability to detect small signals distinguished from noise. More noise implies that weak signals are engulfed by it and lost.

TABLE 8.4

Comparison Table among Microwave Oscilloscope, VNA, and Spectrum Analyzer

Features	Spectrum Analyzer	Vector Network Analyzer (VNA)	Microwave Oscilloscope
Signal characterization	Partial (frequency spectrum)	Not possible	Complete (time domain)
System characterization (LTI systems)	Very limited	Possible (frequency domain)	Possible (time domain)
Nonlinear system characterization (NL-TI systems)	Limited	Not possible	Best option
Sensitivity	High	Best	Poor

8.5 Probe

A measuring instrument is as good as its probe, if one is used. Indeed, the probe forms an integral and vital part of microwave oscilloscopes (Tektronix, 2013b). In general, probes make an electrical connection on the DUT or circuit being measured and carry the signal to be measured to the oscilloscope. Microwave oscilloscopes typically have broadband-fixed 50 Ω SMA terminations, for the probes to be connected. Thus, ideally probes too are required to have large bandwidth. There are many types of probes, depending on the application. For example, some applications may require oscilloscopes to analyze the signal entering a well-terminated system, say, antenna with 50 Ω input. Other applications may demand the oscilloscope to be used to analyze the signal at the transistor drain and source terminals. The former application requires perfect 50 Ω termination when probed by the oscilloscope, whereas the latter application demands that the probe provide ideally infinite impedance. Thus, there are broadly two types of probes used in microwave oscilloscopes. They are:

1. Passive probes
2. Active probes

8.5.1 Passive Probe

A passive probe used in a microwave oscilloscope is as simple as a 50 Ω co-axial cable. These probes are used to analyze the signal across a well-terminated 50 Ω system. For example, we may wish to probe the power amplifier output entering a 50 Ω antenna for signal analysis, like nonlinearity analysis or peak voltage determination. In that case, a well-calibrated 50 Ω co-axial cable is connected between DUT and oscilloscope termination and measurement is made. However, care should be taken regarding the peak voltage levels. Oscilloscopes do indicate a maximum incoming voltage beyond which the damage is severe. In applications where the incoming signal power is significantly high, an attenuator is used to reduce the signal power within oscilloscope limits. Figure 8.21 shows a typical passive probing of a microwave oscilloscope.

Example 8.16

A 10 GHz continuous sinusoidal wave with zero dc offset from a signal generator is required to be probed in a microwave oscilloscope. The signal has a power level of 33 dBm. The absolute peak voltage limit of the oscilloscope is 5 V. It implies that the oscilloscope can sustain a power level 24 dBm for the continuous sinusoidal signal. Hence, in this case a minimum 9 dB attenuator is required before probing the signal.

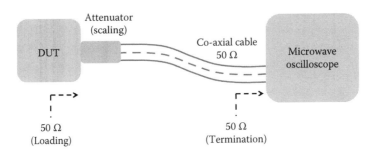

FIGURE 8.21
Microwave oscilloscope: passive probing.

The major limitation of passive probing is loading. In fact, passive probing is designed for probing signals across 50 Ω termination. Hence, they cannot be used to probe the signals across systems with different terminations (other than 50 Ω), or systems that are in operation (for example, the antenna referred to earlier may be permanently fixed to the amplifier feeding it), without loading the system. Passive probing cannot be effectively used to probe the voltage across gate-source (or drain-source) terminals of a transistor or across a diode. Passive probing cannot be used in non-50 Ω systems either. High-impedance active probing addresses the issue of the loading effect seen in passive probing.

8.5.2 High-Impedance Active Probe

As the name implies, high-impedance active probes offer high impedance, and hence present minimal loading effect while probing. Unlike passive probes, the active probes have a buffer amplifier built into the probe. The probe consists of a probe tip or probe head, which may vary depending on application circuit. For example, the probe tip may simply be two sharp needle-like pins, or it may have a co-planar waveguide (CPW) configuration for probing CPW or grounded co-planar waveguide (GCPW) structures. Following the probe tip (or probe head) is the buffer amplifier. The input impedance of the buffer amplifier provides the required high impedance. It is desirable to have high-impedance bandwidth as large as possible. However, in practice the gate-source capacitance limits high-impedance bandwidth. As a result, JFETs (junction field effect transistor) or HEMTs (high electron mobility transistors) are used since they offer low gate-source capacitance, and hence larger bandwidth (Agilent Technologies, 2002, 2011; Tektronix, 2005). Figure 8.22 shows typical block diagram of high-impedance active probing.

High-impedance active probes have become extremely vital for today's advanced communication system design. The ever-increasing demand for high data rates fueled by low-power applications has boosted the role of active devices in microwave communication systems. Active devices demand high-impedance active probes for minimal loading while probing.

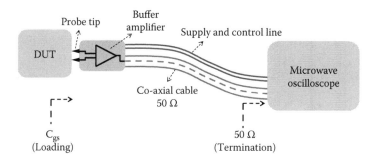

FIGURE 8.22
Microwave oscilloscope: high-impedance active probing.

Continuous rise in bandwidth of real-time oscilloscopes well into the microwave regime has also increased the demand for active probe bandwidth. As mentioned earlier, the gate-source capacitance of FET used in the buffer amplifier is what limits the bandwidth of active probes. Today, it is indeed possible to obtain high-impedance active probes with bandwidths extending up to 30 GHz from dc. It is very important to note that unlike SA (and VNA), microwave oscilloscopes are required to operate right from dc to microwave spectrum. Particularly for active probes, dc behavior is extremely vital. Thus, the low-noise buffer amplifier in the probe is required to have extremely large bandwidth right from dc to microwave frequency. Again, the solution is found in InP and SiGe substrates. The gate-source capacitance in InP transistors is in the order of tens of fF, compared to pF in Si substrate. Figure 8.23 shows a state-of-the-art differential high-impedance active probe with bandwidth up to 30 GHz from Agilent Technologies (2013).

Example 8.17

A 10 fF capacitor offers an impedance of $-j530.5 \ \Omega$ at 30 GHz, as opposed to an impedance of $-j5.3 \ \Omega$ offered by a 1 pF capacitance. This clearly shows how the gate capacitance is vital for enhancing the active probe bandwidth.

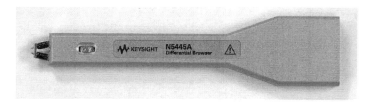

FIGURE 8.23
Agilent active probe. (c) Agilent Technologies, 2014. Reproduced with permission.

High-impedance active probes are not without limitations. Since they have a buffer amplifier implemented in a miniature form factor as the first element, the dynamic range of the probe is even lower than that of the oscilloscope. Any attempt to probe a signal larger than the probe limit may easily damage the active probe permanently. This is a serious limitation of the high-impedance active probe, particularly when used for probing power amplifiers.

High-impedance active probes can further be classified into single-ended probes and differential probes. Single-ended probes are used to probe single-ended voltage signals, i.e., voltage signals with regard to a constant potential node. This constant potential node often is referred to as ground. The ground is common potential for both the DUT and the oscilloscope. It is important to note that the ground of DUT needs to be at the same potential as that of the ground of the oscilloscope for accurate measurement. The diagram in Figure 8.24 clearly depicts the operation of a single-ended probe.

Z1, Z2, and Z3 are parasitic passives like resistance, inductance, and capacitance, which are responsible for making the ground of oscilloscopes different from that of DUT. Thus, in an ideal scenario with an ideal oscilloscope (zero loading) and perfect environment (i.e., Z1, Z2, and Z3 are all zero), just a single probe terminal would have been sufficient. However, in practice

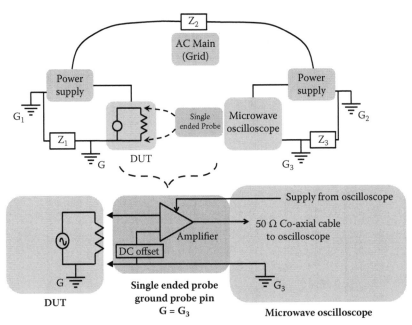

G, G_1, G_2 and G_3 : Commonly referred to as "Ground reference"

FIGURE 8.24
Single-ended probe: a simplified view.

oscilloscopes have finite loading and parasitic, like Z1, Z2, and Z3 make it necessary for probes to have two terminals: one for signal to be measured and another for reference ground contact with minimum impedance path between the ground of the oscilloscope and the ground of DUT. In fact, Z1, Z2, and Z3 are frequency dependent, and particularly at high frequency, even for audio frequency, the parasitics offer significant impedance. Thus, by providing additional ground pin for the probe, the ground of the oscilloscope is locally tied to the ground of DUT with minimum impedance, and hence the measurement is more accurate.

As seen in Example 8.17, the bandwidth of the high-impedance single-ended active probe is mainly determined by the input gate-source capacitance of the buffer amplifier following the probe tip. Hence, it is important to make sure that the probe used for measurement should have higher bandwidth than that of the signal being probed. Otherwise, there will be loading on the circuit being probed (for the harmonics—bear in mind that probing is common for digital systems where the highest frequencies go at least to the third harmonic of the clock), which not only gives incorrect measurement but also affects the state of the circuit being measured, thus making the entire measurement a futile exercise. The general rule of thumb is to use a probe with bandwidth at least three times that of the signal being probed. The bandwidth definition here refers to maximum frequency, where the required high impedance is offered by the probe, so that loading on the circuit being probed is minimum. Again, a rule of thumb is that the probe should offer an impedance at least 10 times the effective impedance across the node being probed. Hence, it may vary from circuit to circuit. Example 8.18 clears this.

Example 8.18

A single-ended active probe has an input capacitance of 0.1 pF.

CASE I

It is used to probe a signal across a 100 Ω load. The probe offers an impedance of $-j1000$ Ω at around 1.6 GHz. Hence, any attempt to probe a signal whose bandwidth is greater than 530 MHz (this is 1600/3) leads to considerable measurement error because of circuit loading.

CASE II

It is used to probe a signal across a 25 Ω load. The probe offers an impedance of $-j250$ Ω at around 6.4 GHz. Hence, any attempt to probe a signal whose bandwidth is greater than 2.1 GHz leads to considerable measurement error because of circuit loading.

Very often the high-impedance active probe is used for measuring switching waveforms like that of high-speed clock signal to measure parameters like rise time, jitter, amplitude noise, and so on. Bandwidth related to rise time becomes very handy in such measurements, not only to validate the accuracy of measurement but also to determine the probe requirement. The following formulae can be used effectively in such situations:

$$B = \frac{0.35}{t_r}$$

(8.34)

$$t_{r\_meas} = \sqrt{t^2_{r\_sig} + t^2_{r\_probe} + t^2_{r\_osc}}$$

(8.35)

where B is 3 dB bandwidth of the effective resistor-capacitor (RC) circuit and t_{r_meas} is the rise time (10% to 90%) of the measured switching signal. Figure 8.25 explains this pictorially by assuming that oscilloscope bandwidth is high enough, and hence its effect is neglected.

Example 8.19

A rising signal with 0.2 ns rise time is to be probed. From Equation (8.34), the 3 dB bandwidth of the signal is 1.75 GHz. Now from the rule of thumb, this signal requires a probe with 3 dB bandwidth at least three times the signal bandwidth. Let us assume that the oscilloscope bandwidth is sufficiently high.

CASE I

Suppose we choose such a probe with required high-impedance 3 dB bandwidth of, say, 6 GHz. In that case, $t_{r\_probe} = 0.058$ ns. Hence, the

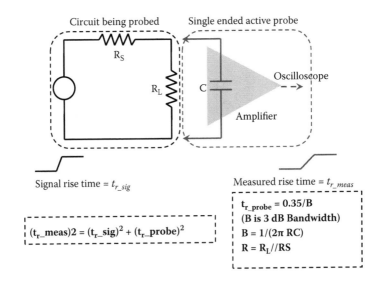

FIGURE 8.25
Switching waveform probed using high-impedance active probe.

measured rise time t_{r_meas} = square root symbol $(0.2^2 + 0.058^2) = 0.208$ ns. This corresponds to an error of 4%.

CASE II

Suppose we choose such a probe with required high-impedance 3 dB bandwidth of, say, 4 GHz. In that case, $t_{r\_probe} = 0.088$ ns. Hence, the measured rise time $v = \text{sqrt} (0.2^2 + 0.088^2) = 0.218$ ns. This corresponds to an error of 9.2%.

CASE III

Suppose we choose such a probe with required high-impedance 3 dB bandwidth of, say, 2 GHz. In that case, $t_{r\_probe} = 0.175$ ns. Hence, the measured rise time $t_{r\_meas} = \text{sqrt} (0.2^2 + 0.175^2) = 0.265$ ns. This corresponds to an error of 32.9%.

From Example 8.19, we see that Equations (8.34) and (8.35) become very handy in applications involving measurement of high-speed switching signals, especially in selecting an appropriate probe for measurement. Another detrimental factor that further limits the probe bandwidth is the grounding inductance. Most often, particularly in high-speed digital circuits, the return current has to loop around if proper a ground path is not provided. Additionally, the probing itself adds a further inductance effect through the probing tip or probing pad. In any case, the effective inductance is in series with the probe's input capacitance. Thus, the probe bandwidth is further deteriorated because of series LC resonance across the load. Figure 8.26 shows a signal of 0.2 ns rise time being probed with different ground inductance values. The probe's input capacitance is 1 pF. Voltage at node A, indicated as V_A, is examined for various possible ground inductances like 0.1, 1, and 3 nH for illustration. It can be seen for the 1 and 3 nH cases that as ground inductance increases, the resonance frequency decreases, which is given by

$$f_r = \frac{1}{2\sqrt{LC}} \tag{8.36}$$

The ringing introduced due to ground inductance not only gives a false signal representation, but also may damage the probe because of overshoot. As seen in Figure 8.26, at 3 nH the ringing is very prominent and not yet settled even after 0.7 ns. Hence, it is extremely important to analyze the possibilities of ground inductance before probing. This may also require careful design of the probe pad (say, with multiple vias), particularly in today's high-speed circuits where switching waveforms have occupied the RF spectrum and made an entry into the microwave regime too. At these frequencies, even the so-thought negligible inductance can be more detrimental in effectively reducing

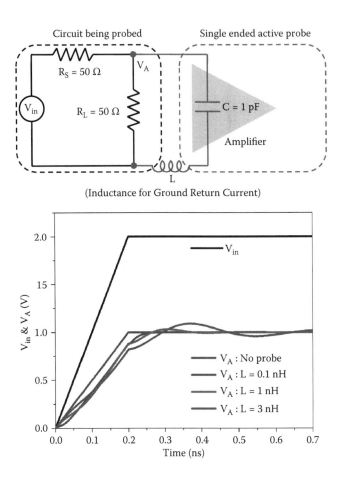

FIGURE 8.26
Ground inductance effect: an illustration.

the probe bandwidth and hence measurement bandwidth, it may also damage it permanently.

Single-ended probes do have their limitations. At microwave frequencies the parasitics are very large (in fact, they are distributed nonstandard components), and any measurement made with a single-ended probe is of questionable value. Of course, single-ended probes cannot be used for measuring differential signals. To deal with differential signals and signal fidelity at microwave frequencies, differential probes are used. In fact, the high-impedance active probes with extremely large bandwidth ranging up to 30 GHz are indeed differential probes. Differential probes use differential amplifiers, and hence can be used to probe differential signals, as shown

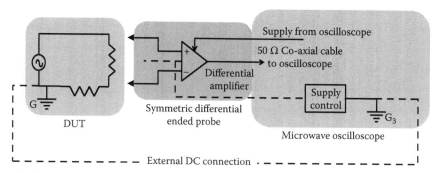

FIGURE 8.27
Differential-ended probe: a simplified view.

in Figure 8.27. In general, differential probes have two symmetrical terminals with no separate terminal for DUT ground reference. However, special
differential probes for battery-operated oscilloscopes do have an additional
terminal for DUT ground reference. Differential signals can also be probed
using two single-ended probes and then using the difference (subtract) function in the oscilloscope. However, there are significant errors that may arise.
First, even slight variation in electrical length between two single-ended
probes while probing transforms into a significant phase difference, thus
leading to skewed and erroneous measurement. In addition, two ports need
to be dedicated for single measurement. Finally, as mentioned, single-ended
probes have higher loading effect and thus lower bandwidth than differential probes. In fact, differential probes have superior performance, and today
they are used even for measuring single-ended signals (Tektronix, 2005;
Agilent Technologies, 2003d).

8.6 Conclusion

An overview of oscilloscope use at RF and microwave frequencies was presented. This instrument has only recently gained popularity at the higher
frequencies, and the microwave versions are still quite expensive, but there
are some applications where they are indispensable. The probe that is often
neglected is actually critical at the higher frequencies, and some details
regarding this were also presented.

Problems

1. A 24 GSps real-time microwave oscilloscope is used to capture carrier and modulated signals from a radio operating at 2.4 GHz. Determine the memory depth required for each of the following (assume each sample is stored in a 1-byte memory location):

   a. Carries signal at 2.4 GHz (sinusoidal signal—capture 10 periods).

   b. Amplitude modulated by 1 MHz square wave (assume ASK modulation and capture 10 envelopes, i.e., 1010101010).

   c. Frequency modulate by 10 MHz square wave (assume FSK modulation with $\Delta f = 100$ kHz). Again, capture the modulated signal for 10 symbols.

   Also determine the acquisition time in each case.

2. A real-time microwave oscilloscope has a sampling frequency of 80 GSps with an equivalent noise bandwidth of 25 GHz. The maximum peak voltage rating of this oscilloscope is 5 V and the port impedance is 50 Ω.

   a. Determine the minimum input power required to be detected and the dynamic range of the oscilloscope.

   b. Compare this with the dynamic range of the spectrum analyzer, which has an intermediate frequency (IF) bandwidth of 100 kHz and peak power rating of 30 dBm across its port.

3. An interleaved ADC architecture using four identical ADCs is shown in Figure 8.28. The individual ADC has a sampling rate of f_0

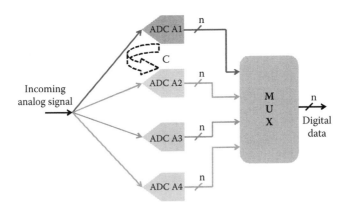

FIGURE 8.28
Four interleaved ADCs, with leakage of input signal.

GSps and a sensitivity (i.e., minimum input signal power required to be detected and sampled) of P_0 watts.

As shown in Figure 8.28, the four ADCs are assembled on a printed circuit board (PCB), one below the other. On measurement following leakage behavior is seen:

Power leaked from ADC A1 = Power leaked from ADC A4 = $(C * P_{in})$ watts

Power leaked from ADC A2 = Power leaked from ADC A3 = $(2C * P_{in})$ watts

Assuming each ADC is equally used, determine the following:

a. Sampling rate of the overall interleaved system of ADC
b. Average sensitivity of the overall interleaved system of ADC (minimum input signal power required to be detected and sampled)

4. In Figure 8.29, determine the following:
   a. Rise time (t_r) when V_s is a step source (t_r is the time taken for V_A to move from 10% V_{A_max} to 90% V_{A_max}, where V_{A_max} is the maximum value of V_A)
   b. Bandwidth (B) when V_s is a sinusoidal source (B is the frequency in Hz at which $|V_A| = 0.707 |V_{A\_max}|$)
   c. Relate t_r and B: Is this same as Equation (8.34)?

5. Determine the rise time for node voltage at A (i.e., for V_A) in each of the circuits shown in Figure 8.30 ($R = 10\ \Omega$, $C = 1$ pF).

Case 1: Calculate rise time using Equation (8.35) in the textbook ($t_r^2 = t_{r1}^2 + t_{r2}^2 + t_{r3}^2 + \ldots$, where t_{r1}, t_{r2}, etc., are individual rise times of each RC circuit in the cascade, whereas t_r is the rise time at node A).

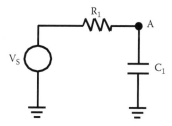

FIGURE 8.29

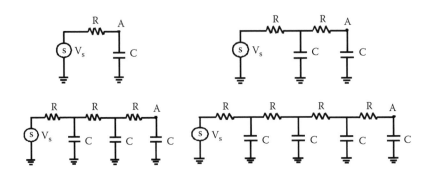

FIGURE 8.30
More complex circuits for timing calculations.

Case 2: Use a schematic simulator and determine the rise time from the simulation (use ideal step input for V_s for simulation).

Case 3: Calculate rise time from simulation like in case 2, but now by using buffers in between cascaded RC sections (use ideal buffer: $S_{11} = 0$, $S_{21} = 1$, $S_{12} = 0$, $S_{22} = 0$).

Compare rise time vs. number of cascades (rise time in Y-axis, number of RC sections in X-axis) for all the three cases.

References

Agilent Technologies. *Agilent Technologies 54701A 2.5-GHz Active Probe*. User Manual. September 2002. http://cp.literature.agilent.com/litweb/pdf/54701-97003.pdf.

Agilent Technologies. *Performance Comparison of Differential and Single-Ended Active Voltage Probes*. Application Note 1419-03. February 2003. http://cp.literature.agilent.com/litweb/pdf/5988-8006EN.pdf.

Agilent Technologies. *Comparison of Measurement Performance between Vector Network Analyzer and TDR Oscilloscope*. White Paper. March 2010. http://cp.literature.agilent.com/litweb/pdf/5990-5446EN.pdf.

Agilent Technologies. *Optimizing Oscilloscope Measurement Accuracy on High-Performance Systems with Agilent Active Probes*. Application Note 1385. May 2011. http://cp.literature.agilent.com/litweb/pdf/5988-5021EN.pdf

Agilent Technologies. *Evaluating Oscilloscope Fundamentals*. Application Note. March 2013a. http://cp.literature.agilent.com/litweb/pdf/5989-8064EN.pdf.

Agilent Technologies. *What Is the Difference between an Equivalent Time Sampling Oscilloscope and a Real-Time Oscilloscope?* Application Note. November 2013b. http://cp.literature.agilent.com/litweb/pdf/5989-8794EN.pdf.

Agilent Technologies. Infiniium 90000 X-Series Oscilloscopes. Data Sheet. November 2013c. http://cp.literature.agilent.com/litweb/pdf/5990-5271EN.pdf.

Agilent Technologies. InfiniiMax III Probing System. Data Sheet. April 2013d. http://cp.literature.agilent.com/litweb/pdf/5990-5653EN.pdf.

Ahmed, I. *Pipelined ADC Design and Enhancement Techniques*. Springer, 2010, Chapter 2. http://www.springer.com/978-90-481-8651-8.

Bhushan, A.S., F. Coppinger, and B. Jalali. Time-Stretched Analog to Digital Conversion. *Electronics Letters*, 34(9), 839–840, 1998.

Bowling, S. *Understanding A/D Converter Performance Specifications*. Application Note AN693. Microchip Technology, April 2000. http://ww1.microchip.com/downloads/en/AppNotes/00693a.pdf.

Coppinger, F., A.S. Bhushan, and B. Jalali. Photonic Time Stretch and Its Application to Analog-to-Digital Conversion. *IEEE Transactions on Microwave Theory and Techniques*, 47(7), 1999.

del Alamo, J.A. Workshop on Advanced Technologies for Next Generation of RFIC. 2005 RFIC Symposium, June 12, 2005.

Jalali, B., F. Coppinger, and A.S. Bhushan. Photonic Time-Stretch Offers Solution to Ultrafast Analog-to-Digital Conversion. *Optics and Photonics News* (Special Issue on Optics), December 1998, pp. 31–32.

Kester, W. High Speed Sampling and High Speed ADCs. Analog Devices, Tutorial. http://www.analog.com/static/imported-files/seminars_webcasts/36701482 52311652750054788237s5ect4.pdf.

Poole, I. Oscilloscope Tutorial and Introduction. Radio-Electronics.com. http://www.radio-electronics.com/info/t_and_m/oscilloscope/tutorial-basics-introduction.php.

Rhode & Schwarz. *Oscilloscope Fundamentals Primer*. Application Note Version 1.1. http://www.rohde-schwarz-scopes.com/primer.php.

Tektronix. *Real-Time versus Equivalent-Time Sampling*. Application Note. January 2001. http://in.tek.com/document/application-note/real-time-versus-equivalent-time-sampling.

Tektronix. *Making Single-Ended Measurements with a Differential Probe*. Application Note. August 2005. http://in.tek.com/document/application-note/making-single-ended-measurements-differential-probe.

Tektronix. Tektronix to Adopt IBM's 9HP SiGe Technology in Next Generation 70GHz Oscilloscopes. News Release. June 2013a. http://www.tek.com/document/news-release/tektronix-adopt-ibm%E2%80%99s-9hp-sige-technology-next-generation-70ghz-oscilloscope-0.

Tektronix. *ABCs of Probes Primer*. Application Note. March 2013b. http://in.tek.com/document/primer/abcs-probes.

Tektronix. *XYZs of Oscilloscopes Primer*. Application Note. February 2014. http://in.tek.com/learning/oscilloscope-tutorial.

Teledyne LeCroy. Teledyne LeCroy Successfully Demonstrates World's First 100 GHz Real-Time Oscilloscope. Press Release. July 2013. http://teledynelecroy.com/pressreleases/document.aspx?news_id = 1801.

9

Wafer Probing

9.1 Overview

In this chapter we will cover some of the important aspects of wafer probing. The term *wafer probing* is almost self-explanatory—it is basically probing the circuits on a semiconductor wafer to measure their electrical performance. The integrated circuit (IC) revolution gave rise to some of the innovative developments across the field of microwave engineering. Concepts that once were thought next to impossible have become a reality today. Concepts have emerged into products. Commercial sophisticated wafer probing systems have not only made their way to numerous research and development laboratories and institutes across the globe, but also become a valued tool in the semiconductor industry as well.

9.1.1 Importance of Wafer Probing

The very first obvious question is: Why should one probe on a wafer? Why cannot one measure the properties of the packaged IC mounted in a printed circuit board (PCB) circuit (this is how it will be used in the end)? This is one of the most important questions because it discloses the reasons that made wafer probing a reality today. Some of these important reasons are as follows:

1. **Device model library creation:** Radio frequency integrated circuit (RFIC) design has become a key sector mainly for the large-volume consumer market. Radio transceivers integrated with system-on-chip (SoC) have dominated the RF domain and are making their way to the microwave domain, even X and K bands, and even into the millimeter-wave domain at the popular 60 GHz (U and V bands).

   A successful RFIC design heavily depends on accurate device models, i.e., transistor models and passives (like inductors, capacitors, transformers, etc.). Wafer probing is the way the device model library is created. In brief, a lot of transistors and other passives are fabricated on wafers covering the range of process parameters and then measured. The measured data are then fitted into the model parameters

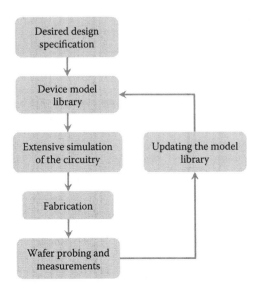

FIGURE 9.1
Typical RFIC design for new product development.

depending on the desired accuracy. Thus, a statistical device model library is developed that then is used in developing a new circuitry for a new product. Since this information will be used to create future ICs, the device information (while the device is part of the silicon wafer) is required and not the packaged component data.

Sometimes designers have their own set of device models derived for their custom IC fabrication process. Figure 9.1 shows an RFIC design flow. The role of wafer probing is also depicted. It can be seen that wafer probing is inevitable for the success of RFIC transceivers, especially today where RFIC radios are being designed at microwave and millimeter-wave frequencies. An accurate device model reduces not only the design and development cost, but also the development time, and hence time to market. With accurate device models the designer can design an optimal IC for a given application with fewer fabrication runs. Thus, wafer probing is necessary for RFIC design.

2. **Cost reduction:** The economics of the semiconductor industry is based on volume production. IC production can sustain only when ICs are fabricated in large numbers and sold at a low unit price. Thus, cost cutting and saving on each product is an important aspect in the IC industry. One such effective cost cutting is while packaging.

Packaging IC die is an expensive process particularly at microwave frequencies. This is because of the sophistication involved in reducing the package parasitics. Traditionally, one would dice out hundreds or thousands of ICs from a single wafer, place each IC in a package, and then characterize the complete component. This would not require wafer probing. However, there are problems associated with it. The important economic problem is that often the yield (successful die in the wafer) of RFIC is less, especially as frequency climbs to the microwave and millimeter-wave regime. Hence, packaging a bad die not only increases the overall package cost per IC, but also increases the overall packaging time and effort or resource. However, if the IC is probed on the wafer itself, then only good (working) die can be sorted, collected, and packaged. This not only helps to reduce overall product cost, but also saves significant time and effort. Thus, wafer probing plays a significant role in sustaining the economy of the semiconductor manufacturing industry.

3. **Achieving optimal system performance:** Wafer probing in addition helps the designer to achieve overall optimal system performance. For example, a package adds significant parasitics, especially at higher frequency. But if the IC can be probed at the wafer level for its true behavior, then the package parasitics (assumed to be known) can be absorbed within the IC design to achieve overall optimal system performance. For example, it may be known that a package lead contributes some known series inductance; now the chip will be designed to achieve the proper capacitance at the terminal so that the inductance is canceled. So the probed data should show the expected capacitance. In fact, the wafer probing setup can itself be used for modeling package parasitics.

Thus, wafer probing adds value not only during new product design, but also during volume production for cost reduction (Breed, 2011; Fisher; Wartenberg, 2003).

9.1.2 Typical Wafer Probing System

A block-level schematic of a typical wafer probing system is depicted in Figure 9.2. Following are the important subblocks of the wafer probing system:

1. Measuring instrument
2. Co-axial cable
3. Wafer probe

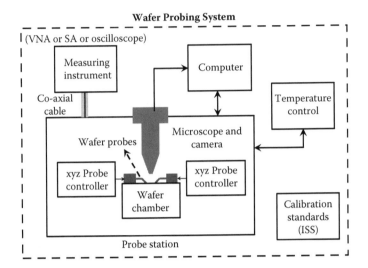

FIGURE 9.2
Block-level schematic of typical wafer probing system.

4. Probe station

5. Calibration standards

1. **Measuring instrument:** The measuring instrument most often used is vector network analyzer (VNA). However, it can be either a spectrum analyzer or a microwave oscilloscope, depending on the application at hand. Wafer probing systems with 110 GHz VNAs are indeed a common presence in most sophisticated laboratories and institutes.

2. **Co-axial cable:** Co-axial cable connects the measuring instrument to the wafer probe. Very often semirigid extremely low loss co-axial cables are preferred for microwave and millimeter-wave measurements.

3. **Wafer probes:** Wafer probes are the distinguishing features of wafer probing. All the sophistication is actually centered around these probes. Figure 9.3 shows a photograph of probes from Cascade Microtech in action (Cascade Microtech, 2010a). Probes launch and capture microwave signals by actually making physical contact to the IC pads. They have extremely small, precisely shaped tips for contact. The tip dimensions are in the order of tens of microns. There are various types of probes for wafer probing, depending on application and performance. We shall discuss the probes in detail later.

4. **Probe station:** The probe station is the mechanical assembly to facilitate proper interfacing and movement of probes and circuits. It includes:

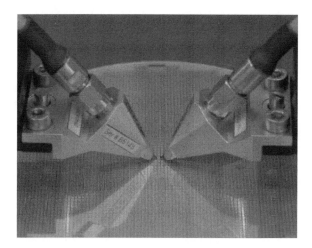

FIGURE 9.3
Photograph of probes in action.

a. **Probe positioner:** The probe positioner holds the probes and positions them accurately on the RFIC pads for probing. The dimensions of IC pads and probe tips are in the order of tens of microns, and hence precise positioning and placing the probe tips on the metals pads is vital. Probe positioners generally come in two configurations:

- **Manual positioners:** Where probes are placed manually by turning the X, Y, and Z knobs on the positioner. They are less expensive and can achieve precise placement of probe tips. They are useful in new product design and development, but not suitable for high-volume wafer probing.

- **Automatic positioners:** They have controllers (computer controlled) that automatically can place the probe tips on the IC pads on a large wafer. They are programmed and controlled by computers with precise prior knowledge of IC pads on a wafer. These automatic positioners are perfect for large-volume wafer probing for IC sorting prior to packaging.

b. **Lighting, microscope, and camera:** An equally important aspect of wafer probing, especially for manual probing. Good lighting and camera make manual wafer probing possible. Without these, there is no manual probing because one cannot place the probes on the IC pads with the naked eye. Even for automatic systems these are important—the system has to be monitored for abnormal situations.

   c. **Wafer chamber:** The chamber is the place where wafer is placed for probing. Hence, the chamber is required to have the following key properties:

     – **Wafer holding by vacuum suction:** The wafer is held in place so that it does not move during probing. At the same time, it should not be held with so much pressure that it breaks. Hence, vacuum suction at controlled air pressure is often used to hold the wafer for probing.

     – **Vibration shielding:** Even the minute vibration caused by human contact with a probe station is sufficient to displace the probe tip from the IC pad. Hence, isolating the wafer chamber from mechanical vibrations is essential. Otherwise, both the probe tip and IC pad may get damaged.

     – **EMI (electromagnetic interference) shielding:** Since the wafer is directly exposed and is being probed, it requires that external EMI be blocked. Especially at millimeter-wave frequency and at hundreds of GHz, the signal of interest is at times a weak signal, and also the ICs at these frequencies can be easily damaged by unexpected voltage/current spikes. Thus, EMI shielding becomes vital as frequency climbs.

5. **Calibration standards:** Calibration is as important as measurement. If the measuring instrument (along with cables and probes) is not well calibrated, then the measurement is useless. Thus, calibration is extremely important, particularly at tens of GHz. There are various calibration standards like SOLT (short-open-load-through), TRL (through reflect line), and LRM (line reflect match) for wafer probing with VNA. Generally, the calibration standards are fabricated on substrates whose properties are precisely known. These substrates are known as impedance standard substrates (ISSs). Seeing the importance of calibration in wafer probing, especially today with rising frequency, we have dedicated an entire section (9.3) to discuss planar calibration standards commonly used in wafer probing.

In addition, a typical wafer probing system would also have the ability to supply dc bias by using dc probes to the IC pads. Furthermore, in large-volume probing for industry products, which most often have integrated system-on-chip (SoC) with a radio, it is required to have the ability to input and probe various digital (or baseband) signals on the IC pads. All these features most often are custom solutions provided depending on the requirement.

A photograph of a typical 110 GHz probe station from Cascade Microtech (2006, 2013a) is shown in Figure 9.4. All the vital subblocks are indentified.

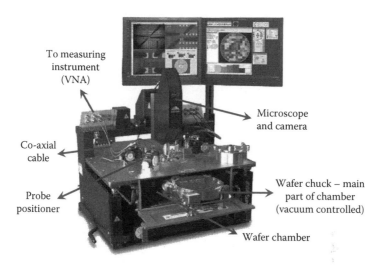

To measuring
instrument
(VNA)

Microscope
and camera

Co-axial
cable

Wafer chuck – main
part of chamber
(vacuum controlled)

Probe
positioner

Wafer chamber

FIGURE 9.4
Photograph of a probe station.

9.2 Commonly Used Probe Configuration

In this section we will discuss some of the commonly used probe configurations, which is one of the most important aspects of wafer probing. As discussed in the overview, most of the sophistication in the system comes from the probes. At microwave and millimeter-wave frequencies, probe parasitics need to be minimized for precise and accurate measurements. Probes make physical contact with the wafer IC pads to launch and capture the microwave signals. They are physically positioned precisely on the wafer IC pads using probe positioners either manually or automatically with computer-controlled motors. There are various types of wafer probes, depending on application. Some of them are rugged for manual probing during design development, and some others are accurate for high-frequency device modeling applications, whereas some are meant for high-powered applications. Apart from microwave probes, there are also probes for feeding dc power supply and digital baseband signals for complete radio and SoC characterization. Figure 9.5 shows a broad classification of wafer probes. In this section we will emphasize the microwave probe configurations (Cascade Microtech, 2013b).

9.2.1 Microwave Probes

A typical schematic and photograph of a microwave probe is shown in Figure 9.6. This photograph of the probe is from Cascade Microtech (Fisher).

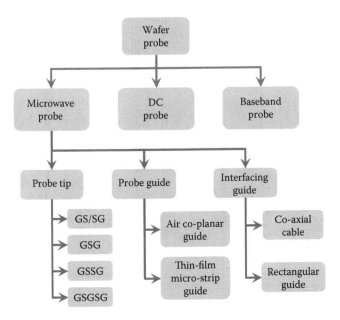

FIGURE 9.5
Classification of wafer probes.

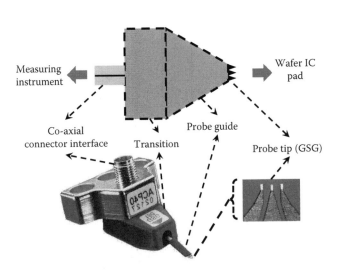

FIGURE 9.6
Schematic and photograph of a typical microwave probe.

The microwave probe typically consists of the following subsections depicted in Figure 9.6:

1. Probe tip
2. Probe guide
3. Interface

The probe tip is that part of probe that makes physical contact with the IC pads. The interface is where the microwave or millimeter-wave signal is available in a standard form (co-axial or waveguide) for connecting to the measuring equipment. The probe guide acts as a guiding medium for the signal from the probe tip to the interface. Let us describe each of these important subblocks of the microwave probe.

1. **Probe tip:** Microwave probes have a characteristic distinguishing feature: they have a multipin probe tip for each signal, unlike dc probes or baseband (used for digital signals) probes. Multiple pins are essential to guide the transmission electron microscope (TEM)/ quasi-TEM signal into/out of the IC pad, especially at microwave frequencies (a common ground does not work). As of today, almost all commercial IC wafer pads are co-planar in nature; i.e., the IC pads for signal and ground all lie in the same plane. Hence, almost all the commercially available and widely used probe tips are also co-planar. The commercial probe tips have several well-known configurations for different applications. Figure 9.7 depicts different probe tip configurations and their applications.

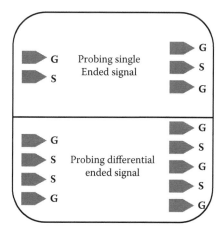

FIGURE 9.7
Schematic of probe tip configurations.

GS/SG is the pioneer in the sense that it is one of the first developed probe tips. It has two pins, one for signal and one for ground. In most cases, the pins are symmetric in construction to support balanced-mode single-ended signals. To probe differential signals from IC, a GSSG configuration is used. In addition, a GSSG probe can also be used for characterizing four port devices and circuits as well. However, as one can predict, both GS/SG and GSSG suffer from signal integrity issues at higher frequencies. GS/SG tends to leak the signals at the probe tip to adjacent IC metal pads, whereas GSSG has significant cross talk between two adjacent signals pins.

GSG and GSGSG tip configurations precisely address these issues. GSG is used for single-ended signals and binds the signal well at the IC probe pads. GSGSG is used for probing differential signals with minimal cross talk, unlike the GSSG configuration. In addition, the GSGSG probe can also be used for characterizing four port devices and circuits, similar to GSSG. The additional ground pin between two signal pins isolates both signal lines, hence reducing cross talk in the GSGSG configuration. However, the price paid for improved performance in GSG and GSGSG is that they require more pads and area as than GS/SG and GSSG, thus increasing the IC area and cost. Most often at millimeter-wave frequencies GSG and GSGSG probe tip configurations are widely used and preferred for their superior performance. Figure 9.8 shows a case study of GSSG and GSGSG probe tip performance by Cascade Microtech (2010b). It clearly shows that GSGSG has superior performance over the GSSG probe tip, especially at higher spectra with reduced cross talk.

2. Probe guide: A probe guide guides the microwave or millimeter-wave signal from the probe tip to a brief length of some standard waveguiding configuration, which in turn connects to the interface. Essentially, the probe tip is extremely small (~100 µm), while external interfaces, such as a co-axial connector, have much larger dimensions—the probe guide is the first step in making this change. Broadly, there are two types of probe guides:

   a. Air co-planar (ACP) guide

   b. Thin-film microstrip guide

   The air co-planar guide supports the co-planar waveguide (CPW) mode of transmission from the probe tip to the interface. Probes with ACP guides are referred to as ACP probes. The ACP probes are the most widely used probes in microwave wafer probing. They are rugged, and more importantly, they are easy to mount (Cascade Microtech, 2010c, 2010d). Hence, most of the manual probing in the microwave regime up to 40 GHz widely uses ACP probes. The probe

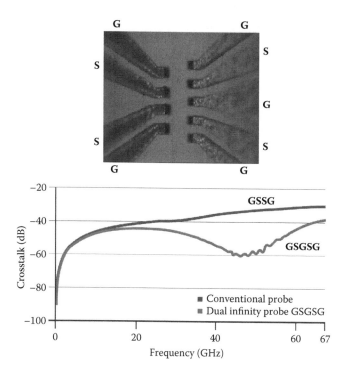

FIGURE 9.8
Cross talk: GSGSG vs. GSSG.

tip extends itself into the CPW guiding medium, as depicted in Figure 9.9 from Cascade Microtech.

However, ACP probes suffer from accuracy issues as the frequency climbs. The main issue is that the signal, while being guided along the probe guide, couples to the wafer and IC pads below. This is because the probe guide is unshielded. Figure 9.9 depicts the coupling nature. This limits the ACP probe's utility at higher frequencies, in particular beyond 40 GHz.

To address this issue of field coupling to IC pads while guiding signals from the probe tip, Cascade Microtech developed a thin-film microstrip-based probe guide. These probes use a microstrip guide, unlike the CPW guide. However, the probe tip is still co-planar since the IC pads are co-planar. Hence, the co-planar probe tip translates to a microstrip probe guide immediately after the IC pad contact region. The solid ground plane shields the probe guide from the IC pads as depicted in Figure 9.10, and hence improves the accuracy of measurement. The thin-film microstrip probe guide has made wafer

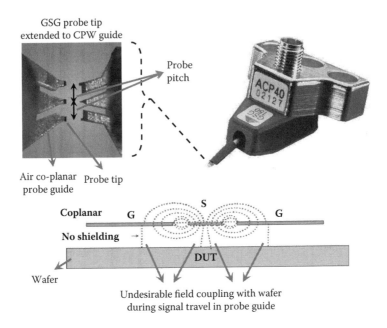

FIGURE 9.9
Air co-planar probe.

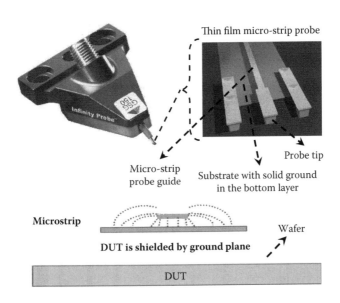

FIGURE 9.10
Thin-film microstrip probe.

probing at 110 GHz and beyond a reality today. Figure 9.10 shows one such microstrip-based wafer probe from Cascade Microtech (Cascade Microtech, 2010d, 2010e, 2010f).

Additionally, thin-film microstrip probes have smaller contact resistance. This further improves the high-frequency behavior of the microstrip probe with lesser parasitics. The schematic of the microstrip probe and the photograph of the probe tip are depicted in Figure 9.11. The only limitation of the microstrip probe is that the probe tip is facing down, as shown in Figure 9.11, and hence placing a probe on the IC pad is not as flexible as the ACP probe. However, today's superior microscope and camera have eased the limitation of probe tip visualization for mounting in microstrip probes. From Figure 9.11, we can also see that contact resistance is reduced by a factor of around four in microstrip probes compared to air co-planar probes. A case study by Cascade Microtech, with the top-view photo of a microstrip probe (also referred to as infinity probe by Cascade Microtech) mounted and removed, is also shown in Figure 9.11, depicting the point contact in microstrip probes (Cascade Microtech, 2010e). Manufacturing of these probe tips and probe guides is really the breakthrough that brought wafer probing into the core of microwave measurements.

3. The interface (waveguide): The interface (also referred to as waveguide, which need not be the rectangular metal waveguide) translates the signal from the probe guide (either CPW or microstrip) to a measuring instrument, which very often is a VNA. Low-loss, high-performance, semirigid co-axial cables are widely used because of their flexibility in handling. However, wafer probes do have the option for rectangular waveguide for >110 GHz in specific high-power or extreme low-loss applications. Figure 9.12 depicts probes with co-axial waveguide as well as rectangular waveguide available from Cascade Microtech (2010f). The type of co-axial cable or the type of rectangular waveguide depends on the frequency of operation.

Before we close this section, let us have a brief look at dc probes and baseband probes. dc probes are used for biasing the transistors as well for power supply in SoC. Generally, dc probes have a single probe pin. The ground is internally tied up inside IC, and hence just a single probe for ground is sufficient. That is, say, to bias a transistor while characterizing, three dc probes are sufficient: one for gate bias, one for drain bias, and one for dc ground at the source of the transistor. However, in applications where dc ground and RF ground are internally tied up inside IC, the third probe for dc ground can also be eliminated. The ground pin of the microwave probe itself will supply the returning dc current for the IC. Figure 9.13 shows a typical dc probe from

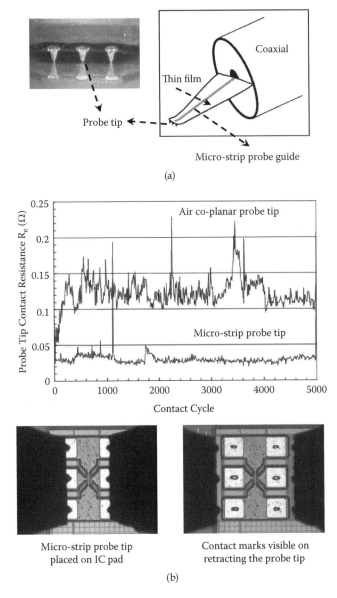

FIGURE 9.11
(a) Thin-film microstrip probe tip. (b) Effects of probe contact.

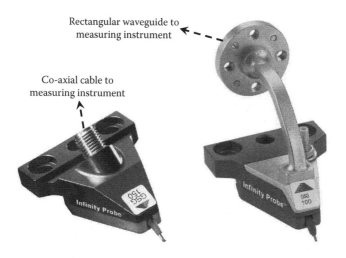

FIGURE 9.12
Waveguide in probes.

Cascade Microtech (2014a). The important aspect in the dc probe is the current density at the contact point. Since the contact dimensions are in the order of microns, it is very important to make complete probe contact at the IC pad to avoid arcing and contact heating, which may damage the IC as well as the dc probe tip. Also, it is advised to make probe contact to the IC pad prior to applying the bias. In special applications for high-powered devices and

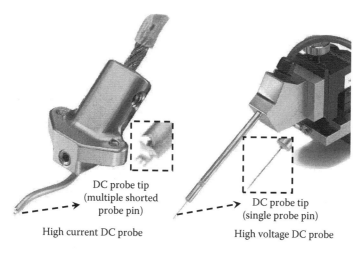

FIGURE 9.13
dc probes.

circuits, specialized dc probes with multiple pins (all connected or shorted) are used to feed the current on multiple adjacent pads to reduce the current density on each pad. Such dc probes are referred to as high-current dc probes. Baseband probes are used to probe the baseband digital signals in the SoC. The frequency of these signals is much less than the microwave and millimeter-wave frequencies. Hence, these probes use a common ground pad for multiple baseband signals. This reduces the chip area, particularly in applications where a radio is integrated on SoC and a complete system needs to be operated and probed. One major concern in baseband probes is the cross talk, since there is no adjacent ground pin to isolate the signals. The baseband probes are available in a variety of forms, depending on requirement. One such probe is depicted in Figure 9.14 from Cascade Microtech (2013b, 2014b).

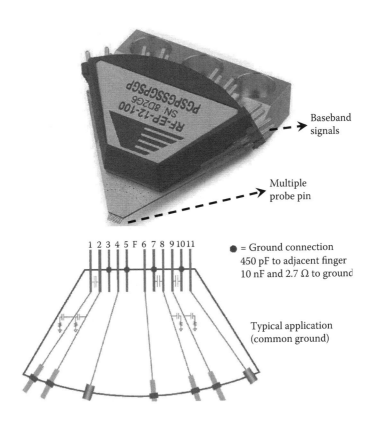

FIGURE 9.14
Baseband probes.

9.3 Planar Calibration Standards

Calibration in general is a vital aspect of microwave measurement, as we have seen earlier. And it becomes inevitable in wafer probing. This is because, as we have seen in the introduction at the beginning of this chapter, one of the main purposes of wafer probing is to extract the device model from vector measurements for RFIC circuit design. An accurate useful device or circuit model can only be extracted by calibrating wafer probes prior to actual measurements. Especially at 100 GHz, even the minute parasitics may hamper overall performance. The parasitic probe capacitance of the order of fF and probe inductance of the order of pH become considerable at millimeter-wave frequencies. As the process technology shrinks, very often the parameter of interest, like drain-gate capacitance, source inductance, etc., is also of the same order as that of the probe parasitics.

Furthermore, the wafer probe itself is quite sophisticated, as seen in the previous section. We have seen the probing pads on the IC are co-planar in nature, and so are the probing tips (either ACP or thin-film microstrip probes). However, the transition from probe tip to the probe guide is no longer co-planar. The CPW mode in the IC pad gets transformed to either air-coupled CPW mode in ACP probes or microstrip mode in thin-film microstrip probes. Additionally, the EM field configuration further gets transformed to either co-axial TEM mode or rectangular waveguide TE_{10} mode in the waveguide of the probe. Thus, in wafer probes often there are more than one field transition. This is depicted in Figure 9.15 for thin-film microstrip probes. Thus, accurate wafer probe calibration is essential in wafer probing.

9.3.1 Planar Calibration Standards

Calibration standards are used to calibrate the wafer probe prior to actual measurements. Since the IC probing pads are co-planar in nature, the calibration standards also have to take the same form where the probe tip contacts them. Some of the commonly used planar calibration techniques in wafer probing are SOLT, SOLR, TRL, LRM, etc. (Cascade Microtech, 2012; Rumiantsev, 2012). Figure 9.16 shows a photograph of the planar LRM standard being used for probe calibration for a GSG probe tip from Cascade Microtech (Fisher). Let us discuss some of these commonly used standards.

9.3.1.1 SOLT and SOLR Standards

SOLT (short-open-load-through) is one of the first and earliest calibration standards. Even today, SOLT is widely used mainly in the lower end of the spectrum. The main advantage of the SOLT standard is that it can be used for broadband probe calibration. Theoretically, with these four standards one

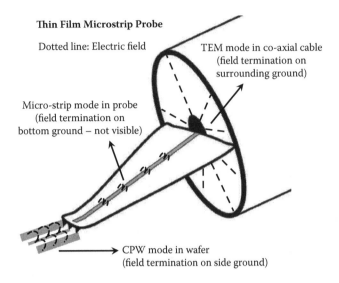

FIGURE 9.15
EM mode variation while wafer probing.

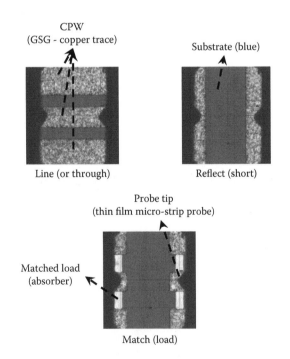

FIGURE 9.16
Planar LRM calibration standards.

can calibrate over any bandwidth of requirement. However, there are practical limits to bandwidth.

But a more important limitation of SOLT calibration in wafer probing is the open calibration. Especially at higher spectra, the open standard suffers from a major limitation. The radiation can become a significant impediment in the open standard, especially at millimeter-wave frequencies. In addition, SOLT calibration requires precise knowledge of behavior of all the standards (SOLT). Hence, the SOLT standard suffers from accuracy issues since a small inaccuracy in each standard can accumulate and increase the overall inaccuracy. Hence, SOLT is used mainly at the lower end of the spectrum until around 20 GHz, mainly for its broadband features and simplicity.

Another important practical problem with planar SOLT standards, particularly in wafer probing, is the through standard. The planar standards are CPW guide based. Hence, the through is required to be straight from probe 1 to probe 2. A practical problem arises when probes are required to be orthogonal, as shown in Figure 9.17. The through standard now bends, which creates unwanted modes at the bend. And SOLT requires precise knowledge of each standard, including the through standard. This problem becomes practically significant in devices under test (DUTs), which are industrial products. To address this issue, SOLT is modified to SOLR (short-open-load-reciprocal) calibration for such applications (Basu and Hayden, 1997). SOLR also has same four standards as SOLT, but the bent through is called the reciprocal standard. However, the major difference is that the reciprocal assumption is made in the SOLR algorithm while calculating the error

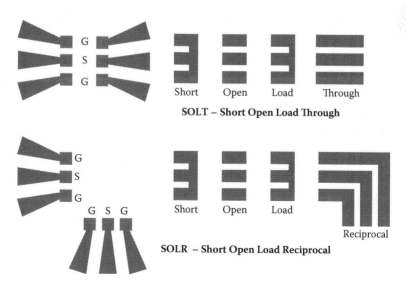

FIGURE 9.17
SOLR and SOLT calibration.

coefficients. It is assumed that the bent through standard is reciprocal, i.e., $S_{21} = S_{12}$, which is an accurate assumption here. This reciprocal assumption relaxes the requirement of precise knowledge of the through as required by SOLT. Hence, SOLR is a viable solution in applications where probes have to be placed orthogonally. But the limitation of the open standard at higher frequency, as seen in SOLT, is also felt in the SOLR standard. Hence, both SOLT and SOLR standards suffer from accuracy issues at the higher end of the spectrum, mainly due to the open standard.

9.3.1.2 TRL and mTRL Standards

TRL (through reflect line) is one of the most widely used and accurate calibrations. TRL is a self-calibration standard in the sense that accurate precise knowledge is not required for any of the standards, unlike SOLT/SOLR. Thus, TRL calibration is accurate even when used at millimeter-wave frequencies. At higher ends of the spectrum, the short is used for the reflect standard in TRL. Thus, the open standard is not required in TRL. The open standard, as we have seen, is the main limiting factor in SOLT/SOLR calibration at higher frequencies due to radiation.

Line standard is similar to through standard with an additional $\lambda_g/4$ length at the center frequency of interest. The beauty of TRL calibration is that it does not require precise knowledge of this additional length, and at the same time, it determines the propagation constant of the planar CPW guide used. The characteristic impedance of the planar CPW guide in the standard becomes the reference impedance of the measurement (it is designed to be close to 50 Ω, but the exact value is important only when ascribing actual values to device impedances, etc., not in the *S*-parameter determination process). Thus, an explicit accurate requirement of 50 Ω load is also relaxed in TRL calibration. Figure 9.18 shows a schematic of TRL calibration standards used.

However, one of the important limitations of TRL calibration is that it is narrowband calibration. As a rule of thumb, TRL is applicable for bandwidths

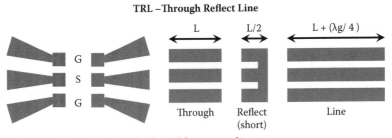

TRL – Through Reflect Line

L   L/2   L + (λg/ 4)

G
S
G

Through   Reflect   Line
(short)

λ_g : Guided wavelength at the desired frequency of interest

FIGURE 9.18
TRL calibration.

below 1:8 (f_l:f_h). The reduced bandwidth is the price paid for improved accuracy in the TRL calibration standard. This is a major limitation especially for device modeling. To address this issue, the National Institute of Standards and Technology (NIST) came up with multiline TRL (mTRL) calibration standards (Marks, 1991). Basically, instead of just one line standard, multiple line standards are used in mTRL. Each line has additional length over the through standard by $\lambda_g/4$ at the various overlapping frequencies to cover the entire band of interest. The entire frequency band is broken down into a number of well-defined frequency bins. Each frequency bin has its line standard. Then all the measurements of overlapping bins are clubbed to get the final broadband calibration. Thus, with the mTRL standard one can have both broadband and accurate calibration combined.

However, mTRL is not without limitations. The limitations are more of feasibility. The first limitation is that multiple line standards at times are not very appealing from the user perspective. For example, to cover a bandwidth from, say, 1 to 100 GHz, the number of multiple line standards can easily exceed five. This not only adds calibration effort, but also increases calibration time. A second limitation of mTRL and TRL in general is that at lower frequencies the line standard can become significantly long. This is because of the additional $\lambda_g/4$ length of line standard over the through standard, which at lower frequencies can easily become too large, and may even exceed the wafer dimension itself. Of course, $\lambda_g/4$ is not essential, as we have seen earlier, and it can be made theoretically much smaller, but this can lead to loss of accuracy. Additionally, the probe positioners also have a limitation on their movement. Hence, at low frequencies line standards can exceed the practical separation between two probes. The LRM standard explained below addresses these limitations of mTRL.

9.3.1.3 LRM

It is very desirable to have broadband calibration like SOLT and accurate like TRL. Line reflect match (LRM) has this combination (Anritsu, 2009; Williams and Marks, 1995). The line in LRM corresponds to a through in TRL, but without any requirement on its length. Again, short is preferred for the reflect standard in LRM as well, and does not require prior characterization. Open tends to be radiative at higher frequencies; hence, short is preferred for the reflect standard. A matched load is used in LRM and determines the reference impedance of measurement. The matched load standard needs to be characterized in advance in LRM, but the important thing is that the matched load need not be perfect over the entire band. An accurate broadband model of the matched load serves the calibration requirements.

The major benefits of LRM are mainly for wafer probing. The user can accurately calibrate the wafer probes over the entire band, say, from 1 to 110 GHz,

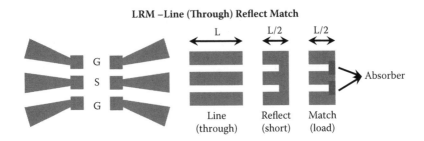

FIGURE 9.19
LRM calibration.

with just three standards in LRM, unlike mTRL, where multiple line standards need to be used. Additionally, all three standards in LRM are smaller in length, and hence perfectly suitable for wafer probing. We saw how TRL suffers from dimension issues mainly at lower frequencies. Thus, LRM planar standards have been extensively used in wafer probing, which achieves accurate broadband calibration and, more importantly, makes them feasible to use. However, it is important to note that LRM requires prior (one-time) characterization of the matched load standard. Figure 9.19 shows the schematic of LRM planar standards.

The above-mentioned standards are some of the commonly used planar calibration standards. However, depending on specific applications and limitations, many versions and modifications of the above-mentioned standards are used in practice. In fact, developing new types of calibration standards for improved accuracy at higher spectra, applicable for specific requirements, is an important research topic.

9.3.2 Off-Wafer and On-Wafer Calibration

Broadly, the wafer calibration can be categorized into two types. Both these methods use the above planar calibration standard discussed in the previous section for calibrating the wafer probes:

1. Off-wafer calibration
2. On-wafer calibration

9.3.2.1 Off-Wafer Calibration

In off-wafer calibration, as the name implies, the calibration is performed on a dedicated substrate with planar calibration standards, and measurement is performed on the wafer of interest. The dedicated calibration substrate is well known as impedance standard substrate (ISS). The ISS has various types of planar standards, like SOLT, TRL, LRM, etc., in planar CPW mode.

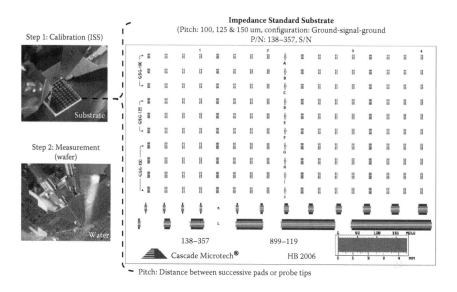

FIGURE 9.20
Off-wafer calibration and ISS.

The ISS has planar standards with varying pitch options and configurations like GS, GSG, etc., so that it can be used for various types of wafer probes. In addition, ISS also has well-defined test structures like capacitors and spiral inductors for verification. Alumina is generally preferred for the ISS substrate. Figure 9.20 shows the ISS used in off-wafer calibration from Cascade Microtech (2014c).

The main advantage of off-wafer calibration is that users can buy the ISS and use it for the calibration. ISS standards are accurately characterized by the vendor. Especially for SOLT and LRM, where prior characterization of the load standard is essential, ISS provides a way.

However, there are limitations of off-wafer calibration as well. Off-wafer calibration is a two-step process. In the first step, the ISS standards are used to calibrate the wafer probe until the probe tip. The probe tips become the plane of reference. Now in the second step the device pads on the measurement wafer need to be de-embedded. This may require identical dummy pads without device for de-embedding the wafer pads and feed lines (or traces). Hence, complete device modeling requires two steps in off-wafer probing. Additionally, there is another subtle limitation. The ISS substrate (most often alumina) may be different from the wafer substrate (most often silicon or GaAs). Hence, the accuracy of calibration is affected due to this substrate mismatch. However, de-embedding the pads on the wafer may reduce this problem to within considerable levels. Figure 9.21 shows the reference planes in off-wafer calibration (Rumiantsev, 2012).

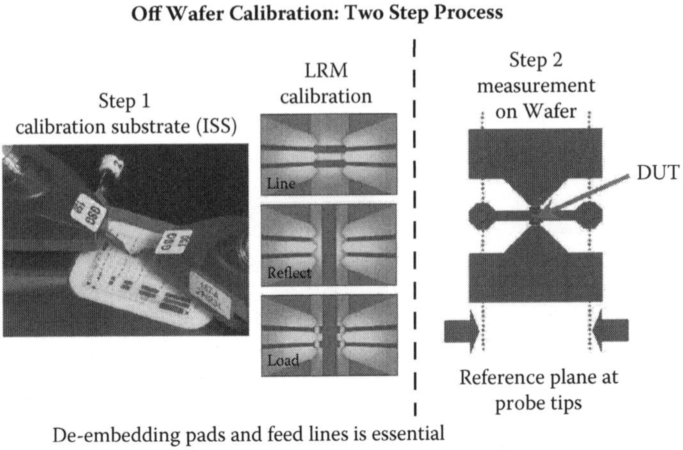

Off Wafer Calibration: Two Step Process

De-embedding pads and feed lines is essential

FIGURE 9.21
Off-wafer calibration reference planes.

Off-wafer calibration is mainly used in applications for device modeling. Because device modeling requires broadband accurate calibration, LRM and SOLT standards are suitable for this requirement. But both of the planar standards require accurate prior characterization of standards. Thus, ISS is suitable in this application where the standards are thoroughly characterized by the vendor.

9.3.2.2 On-Wafer Calibration

In on-wafer calibration, the calibration planar standards are right on the wafer. It does not require an external standard substrate (ISS). The calibration standards are fabricated along with the device on the wafer. Hence, the on-wafer calibration is a single-step measurement. By including the wafer pads and feed lines as part of the custom calibration standards, the reference plane can directly be placed at the device. No additional de-embedding of wafer pads and lines is required since they now are part of calibration standards. Figure 9.22 shows the on-wafer calibration with LRM standards. Observe that pads and feed lines now are part of the custom LRM standards fabricated on the wafer (Rumiantsev, 2012).

The limitation in on-wafer calibration is that the planar standards used for calibration are not accurately characterized at times—without prior characterization of standards, SOLT and LRM become less accurate. Hence, broadband calibration is tough in on-wafer calibration. However, in applications requiring narrowband operation, one can accurately perform on-wafer calibration using TRL standards. In TRL no prior characterization of standards is required. Also, while developing application products, most often radios

On Wafer Calibration: Single Step Process

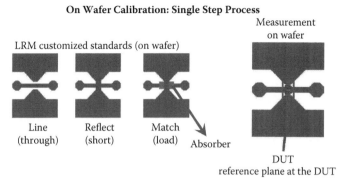

LRM customized standards (on wafer)

Measurement on wafer

Line (through) Reflect (short) Match (load) Absorber

DUT
reference plane at the DUT

Pads and feed lines: Part of custom calibration standards

FIGURE 9.22
On-wafer calibration reference planes.

are narrowband, unlike in device modeling where broadband is required. Hence, in all narrowband applications on-wafer calibration with TRL standards can accurately be used and gives the required measurement right at the device plane with a single step.

References

Anritsu. *LRL/LRM Calibration Theory and Methodology*. Application Note, Version A. March 2009. http://www.anritsu.com/en-AU/Downloads/Application-Notes/Application-Note/DWL3555.aspx.

Basu, S., and L. Hayden. An SOLR Calibration for Accurate Measurement of Orthogonal On-Wafer DUTs. *1997 MTT-S International Microwave Symposium Digest*, Denver, June 1997.

Breed, G. The Basics of Probe Measurements on Wafers and Other Substrates. *High Frequency Electronics*, May 2011.

Cascade Microtech. Summit 6- and 8-Inch RF/Microwave Probing Solutions. 2006. http://www3.imperial.ac.uk/pls/portallive/docs/1/11955696.PDF.

Cascade Microtech. Z Probes Family. 2010a. http://www.cmicro.com/files/z-probe_overview.pdf.

Cascade Microtech. *Making Accurate and Reliable 4-Port On-Wafer Measurements*. Application Note. February 2010b. https://www.cmicro.com/files/EPDAccurate4portAN.pdf.

Cascade Microtech. *ACP Probe Quick Guide*. 2010c. https://www.cmicro.com/files/PGACP.pdf.

Cascade Microtech. *Benefits of Using the New Infinity Probe on Devices with Gold or Aluminum Probe Pads*. Application Note. April 2010d. http://www.cmicro.com/files/INFGOLD_APP_1002.pdf.

Cascade Microtech. Infinity Probe Family. April 2010e. http://www.cmicro.com/files/INFAM_DS.pdf.

Cascade Microtech. *Mechanical Layout Rules for Infinity Probes*. Application Note. April 2010f. http://www.cmicro.com/files/INFINITYRULES_APP.pdf.

Cascade Microtech. *Achieving Consistent Parameter Extraction for Advanced RF Devices*. Application Note. June 2012. https://www.cmicro.com/files/CalParaExtraction_AN.pdf.

Cascade Microtech. Summit 200 mm Semi-Automatic Probe System. Data Sheet. 2013a. https://www.cmicro.com/products/probe-systems/200mm-wafer/summit.

Cascade Microtech. *Probe Selection Guide*. April 2013b. https://www.cmicro.com/files/Probe-Selection-Guide.pdf.

Cascade Microtech. DC Power Probes. http://www.cmicro.com/products/probes/dc-power, 2014a.

Cascade Microtech. RFIC Probes. http://www.cmicro.com/products/probes/rfic-and-multicontact/rfic-and-wedge-probes-for-functional-test, 2014b.

Cascade Microtech. Impedance Standard Substrates. http://www.cmicro.com/products/calibration-tools/impedance-substrate, 2014c.

Fisher, G. *A Guide to Successful on Wafer RF Characterisation*. Cascade Microtech. http://wiki.epfl.ch/carplat/documents/a_guide_to_successful_on_wafer_rf_characterisation.pdf, 2007.

Marks, R.B. A Multiline Method of Network Analyzer Calibration. *IEEE Transactions on Microwave Theory and Techniques*, 39(7), 1205–1215, 1991.

Rumiantsev, A. Establishing S-Parameter Measurement Assurance at the Wafer Level. Cascade Microtech, April 2012.

Wartenberg, S.A. Selected Topics in RF Coplanar Probing. *IEEE Transactions on Microwave Theory and Techniques*, 51(4), 1413–1421, 2003.

Williams, D.F., and R.B. Marks. LRM Probe-Tip Calibrations Using Nonideal Standards. *IEEE Transactions on Microwave Theory and Techniques*, 43(2), 466–469, 1995.

10

Application Examples

In this chapter we will describe two applications of the measurement techniques that have been discussed in the previous chapters.

10.1 Finger Modeling Using Vector Network Analyzer

One of the important aspects of a vector network analyzer (VNA) is in applications for small signal device modeling. In this case study we will try to put forth this application in an interesting way, by deriving an equivalent small signal electrical model of the human finger from VNA measurements. Such an equivalent electrical model of the human finger can potentially be used in capacitive touch sensing applications like track pads in laptops, touch screens in smart phones, capacitive touch buttons, etc. Such an accurate electrical model of the human finger is essential for developing robust touch sensing algorithms and controllers, to minimize and eliminate false detection and true rejection in touch sensors under dynamic environments.

Figure 10.1 shows the measurement setup with VNA. The 50 Ω semirigid co-axial cable has an SMA (M—male-type) connector at one end. The other end of the co-axial cable is connected to the VNA port (say, to port 1). VNA is calibrated using the SOLT (short-open-load-through) calibration standard from 0.4 to 3 GHz until the SMA connector. The finger is then placed on the SMA (M-type) connector, and the reflection coefficient in terms of input S-parameter S_{11} is measured. Since the calibration has been made up to the connector reference, the measured S_{11} can be used directly to derive the electrical model.

To derive the electrical model, let us zoom in at the finger placed on the SMA connector. Figure 10.2(a) shows the schematic of the finger placed on the open end of the SMA connector with a distributed model. Figure 10.2(b) shows an equivalent lumped model of the finger. Now let us see intuitively from where does each part of this electrical model originate. C_{1d} in Figure 10.2(a) is the capacitance between the center pin of the SMA connector (M type) and the outer shield (radio frequency (RF) ground). This distributed capacitance is annular over the entire contact area around the center pin of the SMA connector. It is expected that this capacitance will be lower in value because of the large coupling distance between the center pin and

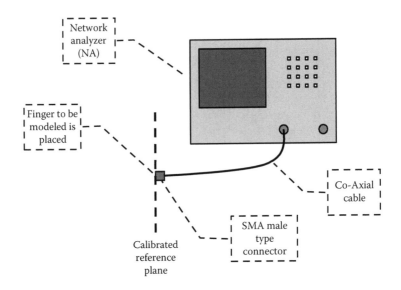

FIGURE 10.1
Measurement setup used in finger modeling.

outer shield. C_1 in Figure 10.2(b) is the effective lumped equivalent capacitance of the distributed counterpart, i.e., C_{1d}. That is, all the distributed small capacitances (C_{1d}) are grouped together into one equivalent capacitor (C_1).

G_{1d} in Figure 10.2(a) is the skin conductance between the center pin of the SMA connector and the outer shield. This distributed conductance is due to the conductive nature of the skin. The conductive nature arises mainly due to salts in skin sweating. The conductance increases in humid environments. The effective lumped equivalent conductance of the distributed G_{1d} is G_1, as shown in Figure 10.2(b). C_{2d} in Figure 10.2(a) is the capacitance between the center pin of the SMA connector and the finger's internal RF ground. The concept of internal RF ground is similar to the ac ground concept of constant

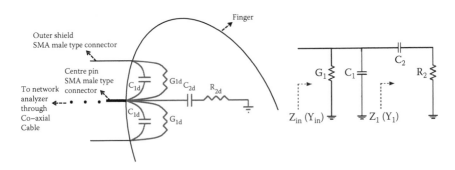

FIGURE 10.2
(a) Physical behavior of finger model. (b) Equivalent lumped finger model.

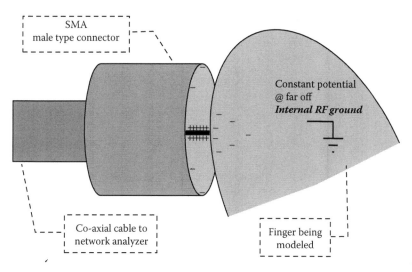

FIGURE 10.3
Concept of internal RF ground in the finger.

potential. The charge in the center pin induces the opposite kind of charge in the finger part that is close to the center pin. However, for the part of the finger that is sufficiently far off, there is hardly any charge induced, and hence it remains at constant potential. Thus, it acts like an RF ground. This is explained pictorially in Figure 10.3. C_2 in Figure 10.2(b) is the effective lumped equivalent capacitance of C_{2d}. It can be expected that C_2 will be much larger than C_1 because of the lower coupling distance.

Finally, R_{2d} in Figure 10.2(a) represents the loss in finger beneath the skin. The induced charge is dissipated as heat by salts in blood and liquid. R_2 in Figure 10.2(b) is the effective lumped equivalent resistance of the distributed counterpart, i.e., R_{2d}. It can be predicted that this value is related to characteristic impedance (Z_0) of the connector.

The next step is to determine the lumped elements, i.e., values of G_1, C_1, C_2, and R_2, from the measured data of S_{11}. The following methodology is one way. First, a close initial value is determined analytically. Then, the final values are determined using a standard well-known curve-fitting method. Commercially available schematic design software can be used for curve fitting. Referring to Figure 10.2(b), the intermediate impedance (Z_1) and admittance (Y_1) are given below:

$$Z_1 = R_2 - \frac{j}{\omega C_2} \tag{10.1a}$$

$$Y_1 = \frac{\omega C_2}{\omega R_2 C_2 - j} \tag{10.1b}$$

ω is the radian frequency that is equal to $2\pi f$, where f is the frequency in hertz. The overall model admittance (Yin) is then derived as

$$Y_{in} = \left[G_1 + \frac{\omega^2 R_2 C_2^2}{1+(\omega R_2 C_2)^2} \right] + j\left[\omega C_1 + \frac{\omega C_2}{1+(\omega R_2 C_2)^2} \right] \qquad (10.2)$$

At low frequency, i.e., when ω is less and hence $\omega R_2 C_2 \ll 1$, (10.2) reduces to following:

$$Y_{in} \approx G_1 + j\omega(C_1 + C_2) \qquad (10.3)$$

From the real and imaginary parts, we can determine the initial values as

$$G_1 = real(Y_{in}) \qquad (10.4a)$$

$$C_1 + C_2 = \frac{imag(Y_{in})}{\omega} \qquad (10.4b)$$

Initially, C_1 is neglected and R_2 is chosen as 50 Ω (Z_0 of the SMA connector used in measurement). The initial value of G_1 is calculated from (10.4a). C_2 is calculated from (10.4b) because C_2 is expected to be much larger than C_1. The initial values determined above are then tweaked in the schematic design software to fit the measurement curve.

The comparison of the equivalent model with the measurement data for one of the fingers (left-hand little finger) is shown in Figure 10.4. The measurement is made up to 3 GHz, which is the upper limit of the particular VNA used for measurement. The Smith chart is the impedance chart with 50 Ω reference. The Smith chart is zoomed at the relevant portion (near high impedance) for clarity. The plot of S_{11} magnitude is also appended in the left top corner. The red solid lines are measured S_{11} data of the finger as obtained using the setup shown in Figure 10.1. The dotted lines are the response of the equivalent electrical model as depicted in Figure 10.2(b), using the component values tabulated in Figure 10.4 for the finger used in measurement. As can be seen, there is a close match between measured and modeled curves. This implies that the proposed model accurately models the finger's electrical behavior from as low as 0.4 GHz to as high as 3 GHz. In reality, for such model to be of practical use, it should be verified rigorously for different people across varying conditions. Such a model with statistical measurement may be of practical significance.

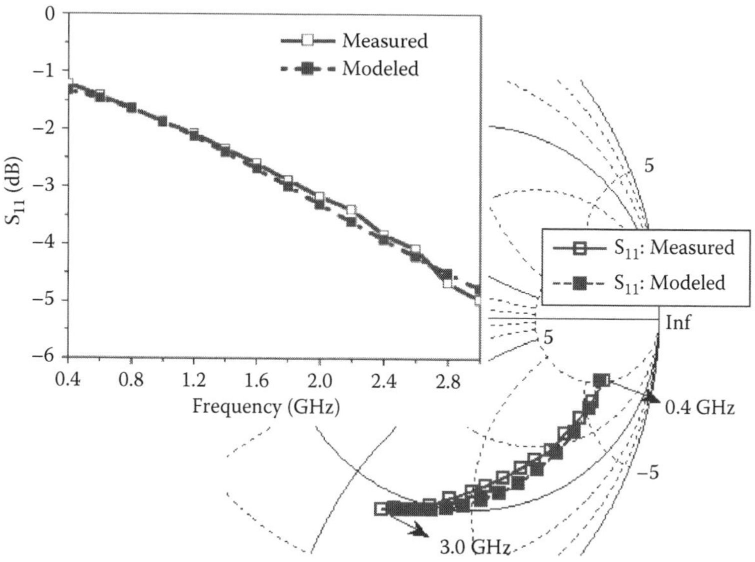

	Left hand little finger			
	G_1 (mS)	C_1 (pF)	R_2 (Ω)	C_2 (pF)
Initial values	1.534	–	50	0.71
Final values	1.399	0.12	58	0.63

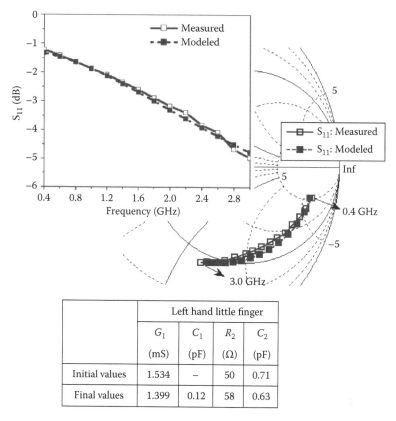

FIGURE 10.4
Measured vs. modeled: left-hand little finger.

10.2 Measurement of Noise Figure of a Mixer

An attempt was made to measure the noise figure of a suspended stripline crossbar mixer (Maas, 1993) operating with an RF and local oscillator (LO) of ~60 GHz, and intermediate frequency (IF) of 0.5 to 2 GHz. The mixer is shown in Figure 10.5.

In the absence of a signal generator at 60 GHz, the LO was supplied from a VNA: LO power was around 0 dBm. The RF signal was obtained by sending a 20 GHz synthesized source output to a Schottky diode-based tripler (filtering of the fundamental is done by the waveguide). Due to the low LO power, the measured conversion loss was 22 dB.

An attempt was made to measure the noise figure of this mixer (at room temperature, which was close to the standard $T_0 = 17°C$) from first principles without using any of the formulae, using a spectrum analyzer. Let us see the

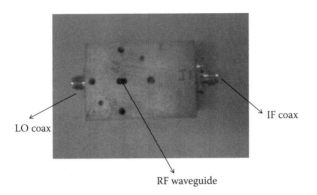

FIGURE 10.5
A crossbar mixer.

steps involved. Throughout, the resolution bandwidth (RBW) of the spectrum analyzer was kept at 10 Hz.

1. Find out the noise generated inside the spectrum analyzer. A 50 Ω termination was connected at the spectrum analyzer input and the noise displayed was checked. The level was -140 dBm $= 10^{-17}$ W. The gain of the spectrum analyzer is assumed to be 1 since a 0 dBm signal input shows on the display as 0 dBm. Now the thermal noise from the termination in 10 Hz bandwidth $= 4 \times 10^{-20}$ W. This is negligible compared to 10^{-17} W. Hence, we can assume that the noise added (N_{sa}) of the spectrum analyzer is 10^{-17} W. *Caution*: It is not easy to estimate the displayed noise level in a spectrum analyzer.

2. A preamplifier of gain $= 24$ dB (or 250) was inserted between the termination and the spectrum analyzer. It was observed that the displayed noise had gone up to -131 dBm (7.9×10^{-17}). Now, $(4 \times 10^{-21} \times 10) G_{amp} + N_{amp} + N_{sa} = 7.9 \times 10^{-17}$. Since G_{amp} and N_{sa} have been already determined, this gives $N_{amp} = 6 \times 10^{-17}$, where N_{amp} is the noise power added by the amplifier to its output.

3. First, measure the noise figure of a 20 dB attenuator to check the procedure since we know that the noise figure (NF) of this component at room temperature is also 20 dB. $G_{Att} = 0.01$.

   When the attenuator was inserted between the termination and the pre-amp, the displayed noise level was -130 dBm ($= 10^{-16}$ W). If the added noise by the attenuator is N_{Att}, then $(4 \times 10^{-21} \times 10) G_{Att} G_{amp} + N_{Att} G_{amp} + N_{amp} + N_{sa} = 10^{-16}$. This gives $N_{Att} = 12 \times 10^{-20}$.

   This gives $NF_{Att} = 1 + 12 \times 10^{-20}/(4 \times 10^{-21} \times 10 \times 0.01) = 301$, or 24.7 dB. We can see that it is in the same order of magnitude (20 dB expected), but not really close. *This is the consequence of reading the displayed noise of a spectrum analyzer by eye estimation and subtracting quantities that are close in value.*

Strictly speaking, the displayed noise should have stayed the same as in the last case (–131 dBm) since a 50 Ω termination by itself and a 50 Ω termination cascaded with a matched attenuator are identical (including noise behavior) as long as the temperature is 290 K. The small 1 dB difference can easily be attributed to estimation error, but it makes a large difference in the final result.

4. Finally, the attenuator was replaced by the mixer (RF and IF ports replacing the attenuator ports). The 50 Ω load was replaced by a waveguide termination. The spectrum analyzer reading was –129 dBm.

Again, we can write the equation $(4 \times 10^{-21} \times 10) \, G_{mix} \, G_{amp} + N_{mix} \, G_{amp} + N_{amp} + N_{sa} = 10^{-15.9}$, where G_{mix} has been measured to be 0.0063 (–22 dB), and N_{mix} is the noise added by the mixer.

This gives $N_{mix} = 2.2 \times 10^{-19}$. This gives the noise figure of the mixer: $\{1 + 2.2 \times 10^{-19} / (4 \times 10^{-21} \times 10 \times 0.0063)\} = 874$, or 29.4 dB.

Again, this is a crude estimate, and from the result for the attenuator, it is likely to be on the higher side.

References

Maas, S. *Microwave Mixers*. Artech House, Norward, MA, 1993.

Appendix 1

Starting from

$$y(t) = \int\limits_{-f_0 - \frac{B}{2}}^{-f_0 + B/2} V_T(f)e^{j2\pi ft}df + \int\limits_{f_0 - \frac{B}{2}}^{f_0 + \frac{B}{2}} V_T(f)e^{j2\pi ft}df \qquad (A.1)$$

assume that $V_T(f) \cong V_T(f_0)$ in the range $f_0 - \frac{B}{2} < f < f_0 + \frac{B}{2}$.
Since $v_T(t)$ is a real voltage signal, $V_T(f) = V_T^*(f_0)$, in the range

$$-f_0 - \frac{B}{2} < f < -f_0 + \frac{B}{2}$$

let

$$y_1(t) = V_T(f_0) \int\limits_{+f_0 - \frac{B}{2}}^{+f_0 + \frac{B}{2}} e^{j2\pi ft}df \qquad (A.2)$$

$$= V_T(f_0) \frac{1}{j2\pi t} \left[e^{j2\pi\left(+f_0 + \frac{B}{2}\right)t} - e^{j2\pi\left(+f_0 - \frac{B}{2}\right)t} \right]$$

$$= V_T(f_0) \frac{1}{\pi t} e^{j2\pi f_0 t} \sin(\pi Bt) \qquad (A.3)$$

Similarly,

$$y_2(t) = V_T^*(f_0) \int\limits_{-f_0 - \frac{B}{2}}^{-f_0 + \frac{B}{2}} e^{j2\pi ft}dt \qquad (A.4)$$

$$= V_T^*(f_0) \frac{1}{\pi t} e^{-j2\pi f_0 t} \sin(\pi Bt) = y_1^*(t)$$

So,

$$y(t) = y_1(t) + y_2(t) = 2Re(y_1(t))$$

If

$$V_T(f_0) = |V_T(f_0)| e^{j\theta}$$

then

$$y(t) = |V_T(f_0)| \frac{2}{\pi t} \cos(2\pi f_0 t + \theta) \sin(\pi B t)$$

Now

$$E = \frac{1}{R} \int\limits_{-\infty}^{\infty} y^2(t) dt = \frac{4}{R} |V_T(f_0)|^2 \int\limits_{-\infty}^{\infty} \cos^2(2\pi f_0 t + \theta) \left(\frac{\sin \pi B t}{\pi t} \right)^2 dt \qquad (A.5)$$

To evaluate this interval, we will use the following well-known relation (Haykin and Vanveen), which is actually a special case of Parseval's theorem

$$\int\limits_{-\infty}^{\infty} |x(t)|^2 \, dt = \int\limits_{-\infty}^{\infty} |X(f)|^2 \, df \qquad (A.6)$$

if $x(t)$ and $X(f)$ are a Fourier transform pair.
   Using (A.6),

$$\int\limits_{-\infty}^{\infty} \cos^2(2\pi f_0 t + \theta) \left[\frac{\sin(\pi B t)}{\pi t} \right]^2 dt = \int\limits_{-\infty}^{\infty} |X(f)|^2 \, df \qquad (A.7)$$

where

$$X(f) = \int\limits_{-\infty}^{\infty} \left\{ \cos(2\pi f_0 t + \theta) \frac{\sin(\pi B t)}{\pi t} \right\} e^{-j2\pi ft} dt$$

$$= \frac{1}{2} \int\limits_{-\infty}^{\infty} \frac{\sin(\pi B t)}{\pi t} \{ e^{-j2\pi(f-f_0)t + j\theta} + e^{-j2\pi(f+f_0)t - j\theta} \} dt \qquad (A.8)$$

This integral can be evaluated if the Fourier transform of $\text{sinc}(t) = \frac{\sin(\pi t)}{\pi t}$ is known.

From well-known tables of Fourier transforms (Haykin and Vanveen),

$$\int_{-\infty}^{\infty} \frac{\sin(\pi t)}{\pi t} e^{-j2\pi ft} dt = \begin{cases} 1 & \text{if } |f| < \dfrac{1}{2} \\ 0 & \text{otherwise} \end{cases} \tag{A.9}$$

Introducing a new variable $u = Bt$, we get the desired relation, which can be used to evaluate (A.8):

$$\int_{-\infty}^{\infty} \frac{\sin(\pi Bt)}{\pi t}(e^{-j2\pi ft})dt = \int_{-\infty}^{\infty} \frac{\sin(\pi u)}{\pi u} e^{-j2\pi \frac{f}{B}u} du = \begin{cases} 1 & \text{if } \left|\dfrac{f}{B}\right| < \dfrac{1}{2}, \text{ i.e., } |f| < \dfrac{B}{2} \\ 0 & \text{otherwise} \end{cases}$$

$$\tag{A.10}$$

So, (A.8) becomes

$$X(f) = X_1(f) + X_2(f) \tag{A.11}$$

where

$$X_1(f) = \begin{cases} \dfrac{1}{2}e^{j\theta} & \text{if } |f - f_0| < \dfrac{B}{2} \\ 0, & \text{otherwise} \end{cases}$$

and

$$X_2(f) = \begin{cases} \dfrac{1}{2}e^{j\theta} & \text{if } |f + f_0| < \dfrac{B}{2} \\ 0, & \text{otherwise} \end{cases}$$

Note that $X_1(f)$ and $X_2(f)$ are nonzero for nonoverlapping ranges of f. So, the products $X_1(f) X_2(f)$, $X_1^*(f) X_2(f)$, and $X_1(f) X_2^*(f)$ are all 0.

Finally, to calculate $\int_{-\infty}^{\infty} |X(f)|^2 \, df = A$, say,

$$A = \int\limits_{-\infty}^{\infty} |X_1(f) + X_2(f)|^2 \, df$$

$$= \int\limits_{-\infty}^{\infty} (X_1(f) + X_2(f))(X_1^*(f) + X_2^*(f)) \, df$$

$$= \int\limits_{-\infty}^{\infty} \left\{ |X_1(f)|^2 + |X_2(f)|^2 \right\} df$$

the other terms being 0, as noted above.

So, $A = \frac{1}{4}B + \frac{1}{4}B = \frac{B}{2}$ from (A.11).

Finally, getting back to (A.5),

$$E = \frac{2}{R} |V_T(f_0)|^2 \, B$$

Index

Printed and bound by CPI Group (UK) Ltd, Croydon, CR0 4YY

22/10/2024

01777621-0015